Chemical Phase Analysis

Chemical Phase Analysis

Roland S. Young

M.Sc., Ph.D., C.Eng., F.R.I.C., F.I.M., F.I.Chem.E., F.I.M.M.
Consulting Chemical Engineer
Victoria, B.C., Canada

Charles Griffin and Company Limited
London and High Wycombe

Charles Griffin & Company Limited

Registered office

5A Crendon Street, High Wycombe, Bucks, HP13 6LE

First published . 1974

ISBN: 0 85264 230 X

SET AND PRINTED IN GREAT BRITAIN BY
LATIMER TREND AND COMPANY LTD PLYMOUTH

Contents

Preface

The differentiation of phases by chemical means has been practised for a long time; a familiar example is the determination of ferrous and ferric iron in most rock analyses throughout the past century. Chemical phase analyses of many elements have been employed in a variety of technologies, but procedures are widely scattered in the literature and are seldom discussed in analytical reference books.

In assembling for the first time in English the published information in this field, and supplementing it with suggestions based on a long experience in phase analyses of extractive metallurgy, the writer hopes this little book may prove useful to those analytical chemists who must differentiate forms from elements.

Victoria, B.C.
January 1974

R. S. YOUNG

Introduction

In many scientific investigations and technical operations it is necessary to know not only the total amount of an element of interest, but also the amount of its different valencies, combinations, or mineralogical forms. The long-established practice of reporting ferrous and ferric iron in rock analyses, the differentiation of copper sulphides from oxidized forms of the element, and the quantitative determination of sulphide, oxide, and metallic phases in many roasting and reducing processes are examples which come readily to mind.

The different phases or forms of an element can usually be observed, of course, by microscopical examination, X-ray diffraction, or other physical means. The frequent difficulty of obtaining by these techniques a rapid quantitative measure of the phases present has led to the development of many chemical procedures, based principally on a selective solvent, a chelate, resin, or other isolating agent. The chemical approach, though it may suffer from certain deficiencies in accuracy and precision, has the merit of quickly providing a reasonably satisfactory result. It does not require expensive and complicated instruments, or workers with long experience in the interpretation of data. The popularity of chemical phase analysis in industrial practice, therefore, is not difficult to understand.

Procedures for the chemical determination of elements in their different phases, combinations, or mineralogical associations are widely scattered in scientific literature. It is hoped that a compilation of these selective methods, together with a critical evaluation and the inclusion of useful manipulative details based on a long experience with many of them, will prove helpful to workers in many disciplines.

The accuracy and precision of chemical phase analyses are influenced by the following variables:

Composition of the sample
Particle size of the sample
Concentration of solvent
Temperature of solution
Agitation of suspension
Atmosphere of reaction
Time of contact of reagents

Composition of the Sample

In phase analyses, a selective-solution step is usually employed to isolate one major phase from the mixture. For instance, in 5% sulphuric acid saturated with sulphur dioxide the oxidized forms of copper such as malachite, azurite, tenorite, and chrysocolla dissolve, whereas sulphide copper is insoluble. The presence of cuprite or native copper, however, complicates such a differentiation because half the cuprite and all the native copper are reported with the sulphides. The presence of copper arsenides or antimonides may pose further uncertainties.

Similarly, it has long been recognized that the familiar determination of ferrous iron in rock analyses may be subject to gross errors in the presence of sulphides, carbonaceous matter, or vanadium.

The sample composition may also affect phase analyses by the presence of substances which intensify or reduce the solvent action through an alteration in oxidation or reduction; examples would be manganese dioxide or metallics in acid solution. Pyrite, for example, increases the solubility of galena in ammonium acetate. The presence of platinum or ferrous sulphate appears to exert a catalytic effect on hydrogen evolution from metallic zinc in acid solution. Conversely, the solvent may react with the sample to form a sparingly soluble compound which coats the particles and slows the reaction rate.

The mineralogical composition or physical form of a sample also has an important bearing on phase analysis. If the element of interest occurs in a finely disseminated state in a hard, impervious siliceous gangue, its extraction by leaching solvents will be more lengthy and less efficient than if that element occurred in the form of large particles interspersed among a soft, porous carbonate gangue.

Particle Size of the Sample

As with all chemical analyses, the representative sample must be finely pulverized to enable the reactions between solvents and solids to proceed to completion within a reasonable period of time. The nature of the material has an important bearing on the degree of comminution required; the determination of less than 1% oxidized copper, occurring in a finely disseminated form in gangue minerals, requires a sample preparation differing from that for the evaluation of metallic iron derived from the reduction of high-grade iron oxide.

Some published methods of phase analyses specify the particle size, but many merely mention a "finely pulverized sample". As a general rule, the sample should be minus 100 mesh; in ores where the element of interest is finely disseminated in gangue, the sample should be minus 200 mesh or even finer.

On the other hand, excessive grinding must be avoided. It is essential, for instance, in a ferrous iron determination to use the coarsest powder that can be decomposed completely by the method of attack employed. Very fine grinding results in the oxidation of ferrous iron, not only due to the greater surface exposure but probably also to localized heating from friction. Contamination with metallic iron and alloying elements from pulverizing equipment is another detrimental effect of over-grinding.

For some phase analyses, the sample does not require to be fine. Examples are the measurement of a metallic phase by the hydrogen evolved on acid treatment, and the separation of metallic aluminium by vaporization in a stream of chlorine or hydrogen chloride.

Concentration of Solvent

The quantity of reagent used to bring about a preferential dissolution of the desired constituent, provided of course that it is adequate for this task, is usually not a critical parameter in phase analyses. In methods where reducing conditions are maintained by a reagent like sulphur dioxide, or oxidizing conditions by hydrogen peroxide, care must be taken that the reagents are fresh enough to provide an excess for the entire reaction period.

Temperature of Solution

Most procedures of phase analyses show a range of solution temperatures from room to about 100°C. The principal reason for elevated temperatures appears to be the increased reaction rate and consequent decreased time required for the determination. Extraction at room temperature or on a water-bath is preferable to boiling, from the viewpoint of supervision, especially if a large number of samples must be handled.

Leaching Time

The time of contact of the selective solvent with the sample may range from a few minutes to many hours. It will depend on several factors: particle size, temperature, agitation, and solubility. The accessibility to the solvent of finely disseminated ore minerals in a gangue of sulphide particles coated with oxide, or of metallic grains surrounded by oxide, is also important in specifying a leaching period. In general, a sample of —100 or —200 mesh size, frequently stirred in a solvent at a temperature about 50°–60°C, will usually not require a leaching period longer than 15–30 minutes. If, however, the desired element occurs in a finely divided state in an impervious gangue, a much longer period may be necessary to attain virtually complete extraction.

Agitation of the Suspension

Some form of agitation to improve solid–liquid contact is usually employed in phase analyses. It may vary from shaking the vessel manually for several seconds every 10–15 minutes during the leaching period, to continuous agitation on a shaking machine for the entire period of the leach.

Two points in the technique of agitation deserve mention. At the outset of leaching, be sure that all particles are thoroughly wetted; large samples of some materials may require considerable swirling of the beaker or flask before complete wetting is achieved. When agitating the suspension, take care that significant quantities of solids are not left on container walls, out of contact with the solution. If necessary, wash down periodically with additional solution.

A vigorous mechanical agitation promotes aeration and oxidation of the sample solution. If this is undesirable in a phase analysis, agitation may be provided by bubbling a stream of nitrogen through the solution.

Atmosphere of Reaction

The differentiation of phases by selective solvents may be carried out in air, or in an inert atmosphere. If the reaction is not sensitive to oxidation, the ease of operation in air dictates its use. When an inert atmosphere is necessary, carbon dioxide is frequently used, perhaps because it can be conveniently generated by an addition of dry ice to the reaction vessel, or of a bicarbonate if the solvent is acidic. Alternatively, an inert atmosphere may be maintained by a flow of nitrogen from a cylinder of the compressed gas.

In some phase analyses, of course, a certain atmosphere is created by the reaction itself, or is induced to promote a separation. For instance, many metals are measured by the hydrogen evolved when the sample is acidified. Aluminium may be separated from its oxide by volatilizing the former in a stream of hydrogen chloride or chlorine.

Limitations of Phase Analyses

The accuracy and precision of phase analyses are usually not as high as those of conventional elemental chemical analyses. Using care in following standard techniques, the precision or reproducibility of phase analyses can be surprisingly good, comparable to that of many routine chemical determinations. The accuracy, or measure of correctness, however, is generally not as satisfactory as that of elemental analysis. In other words, if the true value of metallic iron in a sample of iron and iron oxides is 10·0%, results of replicate phase analyses might lie between 10·4 and 10·6%, with an average of 10·5%.

The accuracy usually varies inversely with the complexity of the material. A differentiation between metallic and oxide zinc in a material containing only these phases would yield results closer to the true values than would a determination of metallic cobalt in a material containing metallic, oxide, and sulphide cobalt together with many compounds of other metals.

Standard samples or pure materials of known composition for checking phase determinations are not as readily available as those used for verifying elemental analyses. Catalogues of chemical reagents usually fail to list Cu_2O, Cu_2S, FeO, Fe_3O_4, sulphides of cobalt and nickel, and several other compounds of interest in phase determinations; in addition, other compounds such as CuS, FeS, ZnS, and oxides of cobalt and nickel are generally available only in a technical grade of unspecified purity. The purity of minerals or compounds used as standards by the analyst should be checked by X-ray diffraction, microscopical examination or other means.

Samples of known composition are invaluable, not only in the development of a suitable procedure for a phase analysis, but also for periodic checking of results in routine work. It is useful to remember that the sum of elements in the separate phases should equal the total quantity of the element in the sample. For example, if a phase analysis indicates that a material contains 7·8% metallic copper, 10·5% cupric oxide, and 12·1% cuprous oxide, a chemical determination for total copper should yield a value very close to 26·9%.

Despite its limitations, chemical phase analysis has provided valuable information in many fields for a long period, particularly in the mining and metallurgical industries. For process control, an approximate result obtained quickly can be very useful; a highly accurate result which becomes available only after considerable time is usually of little value.

Scope of Methods

This review is confined to the quantitative analysis of inorganic phases. Organic compounds, and the qualitative detection of inorganic phases, are not discussed.

Elements of importance in phase analyses are reviewed in alphabetical order. Though emphasis is placed on methods currently in vogue, older procedures are also listed and briefly discussed, because they may not only find application for certain materials but also serve to point out to the analyst other approaches to his problems.

Aluminium

The differentiation of aluminium oxide from metallic aluminium is important in drosses and powders of the aluminium industry, and a number of procedures have been developed to determine these phases. There are five principal methods in use to determine metallic aluminium from alumina.

1. Treatment of the sample with an excess of standard acidified ferric sulphate solution in an inert atmosphere, followed by titration of the ferrous iron produced by metallic aluminium.
2. Reaction of the sample with sodium hydroxide and measurement of the volume or pressure of the hydrogen evolved from the aluminium metal.
3. Passage of dry chlorine or hydrogen chloride at elevated temperature over the sample to volatilize metallic aluminium as $AlCl_3$, leaving alumina as a residue.
4. Treatment of the sample with methanol–bromine mixture to dissolve aluminium metal, leaving the oxide intact.
5. Reaction of the sample with a known solution of a copper salt, filtration of the metallic copper produced by the aluminium metal, and determination of the excess copper left in solution.

These five main procedures will be discussed in detail, followed by brief references to a few other published methods.

1. Ferric Sulphate

The method outlined below is typical of many used for the determination of aluminium in powders having a content of 80–98% [1, 2, 3, 4]. For lower quantities, appropriate modifications may be made in sample weight, strength of oxidant, etc.

From a —100 mesh sample, previously dried at 200°C for 45 minutes, weigh 0·2 g and transfer to a 500-ml wide-necked flask. Carefully add 100·0 ml of ferric sulphate solution so that all powder is washed to the bottom of the flask and none is blown out by pouring the solution directly on the powder. The ferric sulphate solution is prepared by dissolving 330 g $Fe_2(SO_4)_3.9H_2O$ or 235 g of the anhydrous salt in 750 ml of water and 75 ml of sulphuric acid, by stirring and heating, and finally diluting to 1 litre.

Close the flask with a rubber stopper carrying a small tap-funnel and an outlet tube. Immerse the end of the outlet tube in a saturated solution of sodium bicarbonate. Add 50–75 ml of saturated sodium bicarbonate solution through the tap-funnel rapidly, and gently shake the flask. Heat the flask and boil the solution for ten minutes. Cool to below 15°C, remove the stopper, add 15 ml of phosphoric acid, and titrate with 0·5N potassium permanganate solution.

$$\% \text{ Al} = \frac{\text{ml } KMnO_4 \times \text{normality}}{\text{wt. of sample}} \times \frac{26{\cdot}98}{3} \times \frac{100}{1000}$$

If desired, potassium dichromate can be used in place of potassium permanganate; in this case, 2–3 drops of 0·2% solution of sodium diphenylamine sulphonate are added as indicator.

In a modification, ferric chloride is used in place of ferric sulphate, and the resulting ferrous iron is titrated with potassium dichromate [4].

2. Hydrogen Evolution

The procedure described below is representative of those which determine the metallic aluminium in a dross or powder by measurement of the hydrogen evolved when the sample reacts with sodium hydroxide [3, 5].

The reaction vessel is a 300-ml short-necked flask, fitted with a tap-funnel and delivery tube leading to a reservoir bottle of 2–3 litres capacity, containing 250 ml of 1:1 sulphuric acid to trap ammonia evolved from the nitride of the dross. The reservoir, standing in a sink, is connected by means of pressure tubing to a drying tube containing fused calcium chloride or silica gel, and thence to a manometer. The latter consists of 1-mm bore capillary tubing, 400–450mm long, with a safety bulb at its lower end just above the level of the mercury in the outer vessel. A tap above the manometer and drying tube connects the apparatus to a pump.

After testing the apparatus for leaks, weigh 1 g of the –100 mesh sample into the reaction flask, and place 50 ml of 40% sodium hydroxide solution in the tap-funnel. Allow a stream of cold water to flow over the reservoir bottle, open the tap, and pump out the apparatus until the mercury level stands at the top of the manometer, then close the tap. Allow the sodium hydroxide solution to enter the reaction flask, taking care not to admit any air, when the mercury level will fall at once to a point on the manometer tube previously fixed by a blank experiment.

Warm the flask gently to start the reaction and continue to heat until gas evolution slackens; finally boil the solution for about five minutes. The safety bulb on the manometer should prevent any escape of hydrogen.

When the reaction is complete, cool the flask in several changes of cold water, and take the reading of the manometer when it has become steady. The stream of water over the reservoir flask should be maintained continuously during a determination.

The manometer may be calibrated directly in percentage metallics by carrying out the same procedure in duplicate on 0·50 g pure aluminium powder. This gives the mark for 50% metallics, and the scale length between 0 and 50 is then divided into 50 equal parts.

If an appreciable quantity of free silicon or zinc is present, the method is inaccurate unless a correction is introduced for the hydrogen evolved by these metals.

3. Volatilization of $AlCl_3$

The following method is typical of those wherein metallic aluminium is volatilized as $AlCl_3$ in a stream of dry hydrogen chloride, leaving aluminium oxide in the residue [3, 5, 6, 7, 8, 9, 10, 11]. The sample need not be finely pulverized; in fact, a solid piece or several coarse pieces are preferable to fine powder.

Place 0·5–5 g of material in a fused silica combustion boat which has been previously ignited at 1100°C, and weigh the boat and sample. Insert the boat in the inlet end of a Pyrex combustion tube and pass dry hydrogen chloride gas at a rate of about five bubbles per second for 15 minutes. The hydrogen chloride is produced from a generator provided with separate inlets for hydrochloric and sulphuric acids. Oxygen is removed by the addition of hydrogen of about one-tenth the volume of that of hydrogen chloride, and passing the mixture over platinized quartz chippings maintained at a red heat. The gas mixture is dried by passing over calcium chloride, sulphuric acid, and finally anhydrous aluminium chloride.

Raise the temperature of that part of the combustion tube which contains the boat to 400–450°C, and maintain the hydrogen chloride supply for 1–5 hours. Allow the tube to cool, stop the flow of gas, and remove the boat. Ignite the boat and contents in a muffle furnace, cool, and weigh.

If the residue is impure, fuse with potassium hydrogen sulphate, dissolve in a small quantity of hot water, and determine aluminium by any appropriate method.

In a modification of this method, dry, pure chlorine has been used in place of hydrogen chloride, at temperatures from 300° to 550°C [9, 11].

4. Methanol–Bromine

The solubility of metallic aluminium in a mixture of methanol and bromine has been used to differentiate it from alumina [8, 10, 12, 13].

To 0·5–5 g of the —100 mesh sample add 200–300 ml of anhydrous methanol containing 20 ml of bromine. It may be necessary to place the beaker or flask in ice-cold water initially to prevent too vigorous a reaction. Allow the solution to attain room temperature, and finally heat to complete the reaction. Filter, wash with anhydrous methanol, dry the paper and precipitate in a platinum crucible, ignite carefully, and heat in the muffle at 1000°C for 45 minutes. Weigh as Al_2O_3. If the residue is impure, determine the aluminium content by any suitable procedure.

5. Copper Salt

The displacement of copper from cupric sulphate by metallic aluminium can be used to distinguish the latter from alumina [4, 10, 14].

To 0·5–5 g of —100 mesh sample, add a measured quantity of a standard copper sulphate solution; 10–100 ml of 1M $CuSO_4.5H_2O$ acidified with a drop of sulphuric acid will usually be suitable. Dilute with water to about 50 ml if necessary, cover, boil for 15 minutes or until the metallic aluminium has completely dissolved, maintaining the volume at approximately 50 ml by addition of water. Filter and wash with hot water.

The precipitate contains the displaced copper and the aluminium oxide; the filtrate contains the excess copper sulphate and the dissolved aluminium from the metal. Because it is usually easier and faster to determine copper than aluminium, the filtrate is generally analysed directly for copper, or the precipitate is brought into solution by acid treatment or fusion, and copper then determined. Alternatively, or as an additional check, after removal of copper from the above filtrate or precipitate, the aluminium may be determined to give metallic and oxide forms, respectively, of this element. Aluminium may be determined in the presence of copper by atomic absorption using line 3962Å and the nitrous oxide–acetylene flame. Even a large excess of copper has a negligible effect upon the absorption of aluminium.

A modification of this method uses a solution of $CuCl_2.2NH_4Cl.2H_2O$ to dissolve metallic aluminium, leaving the oxide virtually intact [14].

6. Miscellaneous Methods

It has been stated that aluminium oxide in aluminium can be satisfactorily determined by dissolving the sample in 50% hydrochloric acid in a graduated centrifuge tube, and measuring the volume of precipitate formed after centrifuging [15].

In another published procedure, the material was treated with alcohol and mercuric chloride, filtered, ignited, and aluminium determined in the impure alumina of the residue [16].

A novel determination of metallic and oxide aluminium in aluminium bronze has been described. The sample was enclosed in a filter-paper bag placed in a platinum-gauze container, and electrolysed in 100–150 ml of 0·5N sulphuric acid containing 0·5 ml nitric acid, at 2–3 amperes for 3–4 hours until all copper was deposited. Metallic aluminium was determined in the filtrate and alumina in the residue [17].

Metallic aluminium has been determined by adding 10–20 ml of 0·1N iodine solution to a sample containing 10 ml of 2–3N sodium hydroxide, closing the flask for 5 minutes, adding an excess of hydrochloric acid, and back-titrating with standard sodium thiosulphate [18].

Aluminium oxide in aluminium has been determined by the following solution method. Heat 2 g of the finely ground sample in 84 ml of aluminium chloride solution under reflux; the latter solution is prepared by mixing 18 g $AlCl_3.6H_2O$, 75 ml of N HCl, and 0·5 ml of 0·1M $CuCl_2.2H_2O$. Filter the solution and residue through a weighed Gooch crucible, wash, dry, and ignite; the residue is Al_2O_3 [19].

A mixture of sodium tartrate, ammonium sulphate, and ferrous sulphate has been proposed to separate aluminium oxide from the metal, by a rather complicated series of operations [20].

REFERENCES

1. Klinov, I. Y., and Arnol'd, T. I., *J. Applied Chem.* (U.S.S.R.) **9**, 2075–7 (1936). *C.A.* **31**, 2545 (1937).
2. Light, A. K., and Russell, L. E., *Anal. Chem.* **19**, 337–8 (1947).
3. British Aluminium Co., *Chemical Analysis of Aluminium and its Alloys*, London, British Aluminium Co., 1961.
4. Solodovnikov, P. P., *Tr. Kazan. Aviats. Inst.* No. **90**, 64–7 (1966). *C.A.* **68**, 18342 (1968).
5. Furman, N. H. ed., *Scott's Standard Methods of Chemical Analysis*, 6th ed., Vol. 1, Princeton, N.J., Van Nostrand, 1962.
6. Brook, G. B., and Waddington, A. G., *J. Inst. Metals* **61**, Advance Copy No. 772 (1937).
7. Fedorov, A. A., and Linkova, F. V., *Sb. Tr. Tsentr. Nauch.-Issled. Inst. Chernoi Met.* **1962**, 172–8. *C.A.* **58**, 10727 (1963).
8. Bensch, H., *Aluminium* **40**, 743–8 (1964). *C.A.* **64**, 18404 (1966).
9. Fischer, J., and Kraft, G., *Giesserei* **51**, 659–62 (1964). *C.A.* **62**, 3406 (1965).

10. Pimenov, Y. P., *Tr. Mosk. Aviats. Tekhnol. Inst.* No. **67**, 113–21 (1966). *C.A.* **68**, 45976 (1968).
11. Storozhenko, V. N., *Zavod. Lab.* **38**, 1317–19 (1972). *C.A.* **78**, 66599 (1973).
12. Steinhauser, K., *Aluminium* **24**, 176–8 (1942). *C.A.* **37**, 3016 (1943).
13. Werner, O., *Metall.* **4**, 9–12 (1950). *C.A.* **45**, 6536 (1951).
14. Nakamura, T., and Yamazaki, S., *J. Soc. Chem. Ind. Japan* **42**, 296–7 (1939). *C.A.* **34**, 2280 (1940).
15. Klyachko, Y. A., and Gurevich, E. E., *Zavod. Lab.* **6**, 1187–95 (1937). *C.A.* **32**, 1210 (1938).
16. Murach, M. N., Matseev, N. I., Shuikin, N. I., and Makarovskaya, T. A., *Ubileinyi Sbornik Nauch Trudov Inst. Tsvetnykh Metal. i Zolota* **1940**, No. 9, 286–95. *C.A.* **37**, 5924 (1943).
17. Panchenko, G. A., and Remesnikova, E. G., *Zavod. Lab.* **6**, 944–6 (1937). *C.A.* **32**, 454 (1938).
18. Smits, A. A., and Liepina, L. K., *Latviyas PSR Zinatnu Akad. Vestis* **1952**, 103–5. *C.A.* **47**, 8578 (1953).
19. Boner, J. E., *Helv. Chim. Acta* **28**, 352–61 (1945). *C.A.* **40**, 533 (1946).
20. Denisov, E. I., *Trudy Leningrad Politekh. Inst.* **1959**, No. 201, 102–9. *C.A.* **54**, 11823 (1960).

Antimony

Phase analysis of antimony compounds has been confined to the differentiation of oxides from sulphides, and of the trioxide from the pentoxide, in minerals, ores, concentrates, calcines, and flue dusts.

1. Oxides and Sulphides

To 0·5–2 g of –100 mesh sample, add 50–100 ml tartaric acid and 5 g of potassium acid tartrate. Boil gently for 2 hours, maintaining the original volume by the addition of water. Filter through a weighed Gooch or sintered glass crucible, wash with hot 5% tartaric acid, dry at 110°C in an oven for an hour, and weigh the antimony sulphide. The latter will almost invariably be stibnite, Sb_2S_3. The oxides of antimony are dissolved in tartaric acid–tartrate solution, and pass into the filtrate where the antimony content can be determined by any suitable method [1, 2].

2. Differentiation of Oxides

Antimony trioxide in a flotation concentrate can be separated from the pentoxide by boiling 1 g of the –100 mesh sample with 100 ml of 5% tartaric acid for 25 minutes, filtering, and washing with hot water. The trioxide dissolves and passes into the filtrate, whereas Sb_2O_5 is insoluble in 5% tartaric acid and remains as a residue [3].

It has been reported that the minerals valantinite, Sb_2O_3, and hydroservantite, $Sb_2O_3.Sb_2O_5$, can be separated by the solvent 1.5N tartaric acid for the former and 0·8N sodium sulphide for the latter, from hydromeite and from the sulphide mineral antimonite. The latter can be determined indirectly through the sulphur content, and hydromeite is then obtained by difference [4].

REFERENCES

1. Lebedev, S., and Mikie-Indin, K., *Zbornik Radova Srpska Akad. Nauk* No. **22**, Geol. Inst. No. 2, 169–76 (1952). *C.A.* **47**, 6820 (1953).
2. De Vadder, P., *Chim. anal.* **35**, 248–50 (1953). *C.A.* **48**, 75 (1954).
3. Gayupov, G. R., and Ibragimov, Y. I., *Uzb. Khim. Zh.* **16**, (2), 17–18 (1972). *C.A.* 77, 69692 (1972).
4. Solntsev, N. I., and Dubovitskaya, E. I., *Analiz Rud Tsvetnykh Metal. i Produktov ikh Pererabotki* **1956**, No. 12, 24–35. *C.A.* **51**, 14484 (1957).

Arsenic

A few papers have described phase analyses of arsenic compounds in ores, flue dusts, and reagents.

Arsenic pentoxide in arsenic trioxide can be readily determined by treating with potassium iodide in the presence of strong hydrochloric acid and titrating the liberated iodine with standard sodium thiosulphate. The presence of Fe, Cu, and Sb interfere, and it is advantageous to remove As_2O_5 with water, in which it is more soluble than are the other oxides [1].

A simple method of determining free arsenic trioxide in calcium arsenite consists in volatilizing the former at 200–250°C; calcium arsenite is chemically stable at 300°C [2].

By heating a sample of arsenic ore with Al_2O_3 in a H_2O–CO_2–HCl atmosphere at 500–550°C, arsenido-sulphides decompose to yield sulphur as H_2S and arsenic as $AsCl_3$, while arsenites are volatilized as $AsCl_3$, and sulphoarsenates give H_2S and As_2O_5. Arsenates yield As_2O_5. Arsenides are separated from As_2O_5 by the solubility of the latter in hydrochloric acid [3].

Another scheme for phase analysis of arsenic compounds in flue dusts has been published; it includes oxides, arsenates of lead and zinc, sulphides, and metallic arsenic. Unfortunately, no details are obtainable from the abstract [4].

REFERENCES

1. Lambie, D. A., *Analyst* **73**, 74–8 (1948).
2. Artamonov, N. S., and Bakhtiarova, Z. K., *Zavod. Lab.* **5**, 1176–9 (1936). *C.A.* **31**, 1726 (1937).
3. Chernyi, A. T., and Podoinikova, K. V., *Zavod. Lab.* **16**, 1031–5 (1950). *C.A.* **45**, 1913 (1951).
4. Filippova, N. A., *Analiz Rud Tsvetnykh Metal. i Produktov ikh Pererabotki* **1956**, No. 12, 99–100. *C.A.* **51**, 14473 (1957).

Beryllium

Very little work has been done on the phase analysis of beryllium.

Beryllium oxide in beryllium metal has been determined by solution of the latter in methanol–bromine, similar to the reaction for aluminium. To 0·5–5 g of –100 mesh material add 200–300 ml of anhydrous methanol containing 20 ml of bromine. Allow to react at room temperature and finally heat until only the residue of beryllium oxide remains. Filter, wash with anhydrous methanol, dry the paper and precipitate in a platinum crucible, ignite carefully, and heat in the muffle at 1000°C for 45 minutes. Weigh as BeO. If the residue is impure, determine the beryllium content by any convenient procedure. It may be necessary with some materials to complete the solution of beryllium metal by addition of methanol–hydrochloric acid after the reaction with methanol–bromine has proceeded to completion [1].

A method has been published for determining beryllium oxide in beryllium carbide by selective bromination of the latter [2]. When bromine is passed over the sample of beryllium carbide heated at 825°C, beryllium bromide is formed which can be swept out of the apparatus by argon. Beryllium oxide is left in the combustion boat and can be weighed.

The unusual property of beryllium oxide, of being insoluble in molten sodium carbonate, has apparently not been utilized in phase analyses, but deserves to be borne in mind by chemists working with this element.

REFERENCES

1. Eberle, A. R., and Lerner, M. F., *Metallurgia* **59**, 49–52 (1959).
2. Reed, S. A., Funston, E. S., and Bridges, W. L., *Anal. Chim. Acta* **10**, 429–42 (1954).

Bismuth

Several workers have investigated the differentiation of metallic, oxide, and sulphide bismuth.

Metallic bismuth in the presence of bismuth oxide can be readily estimated [1]. To 0·5–1 g of –100 mesh sample add 25 ml of 1N silver nitrate solution, dilute with water to approximately 50 ml, and allow to stand about 15 minutes, shaking or swirling frequently. Filter, wash thoroughly with water, and titrate in the filtrate the excess silver nitrate with 0·1N sodium chloride in the presence of potassium chromate as indicator. Only the metallic bismuth reacts with silver nitrate.

In another method, bismuth oxides are selectively dissolved in dilute hydrochloric acid + phenyltartaric acid at 80°C, followed by dissolution of metallic bismuth from the precipitate in dilute nitric acid and silver nitrate at room temperature; bismuth sulphide remains in the final residue [2].

A different approach has been used in other procedures. In one, oxide minerals are dissolved by a solution of 5% thiourea in 0·5N sulphuric acid, previously de-aerated by hydrogen or nitrogen, for one hour [3]. Filtration leaves a residue from which metallic bismuth is removed by stirring for 30 minutes with 0·1N silver nitrate in 0·5N nitric acid. Filtration of this suspension leaves sulphide bismuth in the final residue, to be determined by any convenient method. Metallic bismuth = total bismuth – (oxide + sulphide bismuth).

In another contribution, metallic bismuth is leached with 0·1N silver nitrate and 0·3N nitric acid for one hour; in another sample metallic bismuth and the trioxide are leached by 0·5N nitric acid [4].

REFERENCES

1. LONG, L. H., and SACKMAN, J. F., *Research* **7**, S17–18 (1954).
2. BOGATYREVA, L. P., and AFANAS'EVA, G. V., *Zavod. Lab.* **36**, 916–18 (1970). *C.A.* **73**, 137039 (1971).
3. MERLINA, F. E., and BUDNIKOVA, N. V., *Zavod. Lab.* **35**, 1169–71 (1969). *C.A.* **72**, 50552 (1970).
4. FILIPPOVA, N. A., and KOROSTELEVA, V. A., *Zavod. Lab.* **30**, 518–22 (1964). *C.A.* **61**, 4963 (1964).

Boron

The chemical phase analysis of boride, carbide, and borocarbide mixtures has been reported [1], but no details are available from the abstract.

REFERENCE

1. VEKSHINA, N. V., and MARKOVSKII, L. Y., *Khim. Svoistva Metody Anal. Tugoplavkikh Soedin*, **1969**, 132–42. *C.A.* **72**, 62502 (1970).

Cadmium

The forms of cadmium in flue dusts and other metallurgical materials can be separated [1, 2]. Cadmium sulphate is first extracted with water from the —100 mesh sample. Then a 2-hour leach of the residue with 5 g/l sulphuric acid at room temperature with occasional stirring, dissolves cadmium oxide, together with the compounds n $CdO.CdSO_4$, $CdO.SiO_2$, and 2 $CdO.SiO_2$. Treatment of the residue for 2 hours at 90–95°C with 10 g/l copper sulphate and 5 g/l sulphuric acid dissolves metallic cadmium, cadmium sulphide, and the compound $CdO.Al_2O_3$. A further leach of the resulting residue with 50 g/l sulphuric acid for 2 hours at 90–95°C dissolves the compound $CdO.Fe_2O_3$, leaving in the final residue such difficultly-soluble substances as n CdO.m $ZnO.Fe_2O_3$.

REFERENCES

1. Tseft, A. L., and Kabanova, L. M., *Trudy Altaisk. Gorno-Met. Nauch.-Issledovatel. Inst. Akad. Nauk Kazakh. S.S.R.* **1**, 79–86 (1954). *C.A.* **51**, 14480 (1957).
2. Sudilovskaya, E. M., and Filippova, N. A., *Analiz Rud Tsvetnykh Metal. i Produktov ikh Pererabotki, Sbornik Nauch. Trudov* **1958**, No. 14, 138–42. *C.A.* **53**, 13887 (1959).

Calcium

1. Free Calcium Oxide

In many industrial operations, large quantities of quicklime or hydrated lime are used in the form of dilute aqueous solutions or suspensions for *p*H control or as a source of calcium ions. The term "free lime" or "available lime" is used to denote the actual content of uncombined calcium oxide in these products. It gives a measure of the quality of the lime, because total calcium oxide would include unburned limestone which is insoluble in water and therefore of no value for milk-of-lime. There are several methods used to obtain the free or available lime in quicklime or hydrated lime.

A simple but effective differentiation of calcium oxide from unburned limestone, long used in the field of extractive metallurgy, is based on the fact that calcium oxide forms with a sucrose solution a water-soluble saccharate which can be readily titrated with standard oxalic acid solution, whereas calcium in the form of carbonate and silicate does not react with sucrose [1]. Place 1 g of —100 mesh material in a 500 ml volumetric flask, add 30 g of sugar and 300 ml of water; shake vigorously until the sugar has dissolved. Make up to 500 ml and shake several times at 10-minute intervals for an hour. Filter a portion of the solution, and withdraw 50 ml of this filtrate for titration. Add 2 drops of phenolphthalein and titrate with 0·1N oxalic acid to the disappearance of the pink colour. One ml of 0·1N oxalic acid = 0·0028 g CaO. The 0·1N oxalic acid is made by dissolving 6·303 g $H_2C_2O_4.2H_2O$ in water, and diluting to 1 litre.

It has also been demonstrated that this method can be used in the presence of the phosphate, fluoride, and sulphate of calcium, and of the carbonate and oxide of magnesium [2].

Another method used in industrial practice for available lime has been described [1]. Weigh 1·402 g of —100 mesh material into a 400-ml beaker, add 200 ml hot water, and boil for 3 minutes. Cool, add 2 drops phenolphthalein and titrate with 1·0N hydrochloric acid. When the pink colour disappears in streaks, reduce the rate of acid addition until the colour fades throughout the solution for 1–2 seconds. Note the reading and ignore the return of colour. Repeat the experiment, substituting for the 400-ml beaker a 1-litre volumetric flask carrying a one-hole stopper fitted with a short glass tube drawn out to a point. Cool, and add dropwise with vigorous stirring about 4–5 ml less acid than before. Call this number of ml used "A". Dilute to the mark with freshly boiled distilled water, close the flask with a solid stopper, mix well for 4–5 minutes and allow to settle for 30 minutes. Pipette off a 200 ml portion, add phenolphthalein, and titrate slowly with 0·5N hydrochloric acid until the solution remains colourless on standing 1 minute. Call this additional number of ml "B". Then the percentage of available CaO = 2A + 5B.

Uncombined calcium oxide in lime and silicate products may be extracted with boiling glycerol–alcohol and titrated with ammonium acetate in ethanol, using phenolphthalein indicator; an accelerator such as strontium nitrate should be added [3]. The same solution can also be titrated with hydrochloric or sulphuric acid, using methyl red indicator; conductometric titration of glycerol–alcohol or ethylene glycol may also be employed. Conductivity measurements are not suitable where the sample is extracted with glycerol in the presence of an accelerator. Best results are obtained by conductometric titration with a strong acid. Reasonable values can also be secured using an indicator: methyl red for glycerol and ethylene glycol extracts, and bromophenol blue for acetoacetic extracts.

In another variation with non-aqueous solvents, calcium oxide is extracted with glycerol, ethylene glycol, or phenol, and titrated with EDTA at *p*H 10 obtained with an ammonia–ammonium chloride buffer, using Eriochrome Black T indicator. Titration may also be carried out with 0·05N ethanolic hydrochloric acid, with 3 drops alizarin S and 2 drops of bromocresol green as indicator [4].

A method for free calcium oxide in calcium borates has been published [5]. After drying the —100 mesh material at 100°C for 8–10 hours, and heating at 600°C for 2 hours, reflux it with a mixture of absolute ethanol and absolute glycerol. Carry out a titration with 0·1N benzoic acid, using phenolphthalein indicator.

Free calcium oxide can be extracted from the carbonate of calcium and magnesium by heating the —100 mesh sample for 15–20 minutes with 5% ammonium chloride solution, filtering, and titrating with EDTA [6].

2. Calcium Phosphate in Carbonates

Calcite, aragonite, and dolomite can be separated from calcium phosphate by the following method [7]. Digest 1 g of the —200 mesh sample in 100 ml of 0·5M triammonium citrate solution at 65°C for 4 hours with constant agitation. Allow the mixture to stand at room temperature for 18 hours, filter, wash, and dry the residue of calcium phosphate; carbonates are in the filtrate.

REFERENCES

1. Young, R. S., *Chemical Analysis in Extractive Metallurgy*, London, Charles Griffin, 1971.
2. Young, R. S., *Talanta* **20**, 891–2 (1973).
3. Assarson, G. O., and Bokstrom, J. M., *Anal. Chem.* **25**, 1844–8 (1953).
4. Verma, M. R., Bhuchar, V. M., Therattil, K. J., and Sharma, S. S., *Analyst* **83**, 160–8 (1958).
5. Burnazyan, A. S., Karibyan, A. N., and Simanyan, I. M., *Arm. Khim. Zh.* **22**, 215–18 (1969). *C.A.* **71**, 67148 (1969).
6. Alferova, V. N., and Valyuzhinich, R. N., *Khim. Prom. Ukr.* **3**, 49–51 (1968). *C.A.* **70**, 8653 (1969).
7. Smith, J. P., and Lehr, J. R., *J. Agr. Food Chem.* **14**, 342–9 (1966).

Carbon

1. Graphite

In cast iron, part of the total carbon in the metal is present as graphite, and the determination of the latter has long been practised in the iron and steel industry. The method outlined below is typical [1].

Dissolve 1–3 g in 50 ml of nitric acid on the steam bath. Add a few drops of hydrofluoric acid to facilitate the filtration of any separated silicic acid. Filter through ignited asbestos on a small Gooch crucible which will fit in the combustion tube of the carbon train or apparatus. Wash thoroughly with hot water, hot 12% potassium hydroxide, hot water, 5% hydrochloric acid, and finally with hot water. Dry at a temperature not exceeding 150°C, and determine the graphite by direct combustion in oxygen at approximately 900°C in the apparatus used for the gravimetric determination of total carbon. The combustion tube must be closed immediately after inserting the sample. If desired, the evolved carbon dioxide from the combustion of graphite may be measured volumetrically, conductometrically, by low-pressure combustion, thermal conductivity, or infrared analyser, in place of the conventional gravimetric technique.

2. Industrial Diamond

Diamonds in industry are usually held in a matrix of metal, ceramic, cemented carbide, plastic, or rubber; diamond polishing powders may be incorporated in a variety of pastes. The determination of diamond therefore involves a separation from the other forms of carbon which are almost invariably present [2].

As a general rule, the determination of the diamond content of any material must be accompanied by a report on the size of the diamond particles. Consequently the crushing and pulverizing operations to reduce the size of the diamond-containing material should be kept to a minimum consistent with securing a representative portion for analysis.

To the sample in a beaker add nitric, hydrochloric, and sulphuric acids and evaporate to fumes of the latter. For materials high in organic matter, repeated evaporations with additional nitric acid may be required. Transfer the sample to a large platinum dish, add nitric and hydrofluoric acids, and digest on the hot plate until all silica has been volatilized and complete decomposition of the material has occurred.

Carefully decant the solution from the residue of diamond, and wash several times by decantation. If a yellow precipitate of tungstic oxide has been formed by evaporating the solution too far, dissolve this in ammonium hydroxide before separating the diamonds by decantation. If particles of refractory carbides are still present with the diamond residue, fuse the latter with sodium bisulphate in a Pyrex "copper" flask, or with potassium hydroxide in an iron or nickel crucible. Solution of the cooled melt in hot water leaves the diamond, with traces of other carbonaceous material, as a residue. Transfer the latter to a beaker, and boil thoroughly with a saturated solution of sodium dichromate in sulphuric acid to eliminate non-diamond carbonaceous residue. Filter through a fine porosity sintered glass filter and weigh the diamond residue.

3. Free Carbon in Refractory Compounds

Total carbon in cemented carbides is determined by conventional combustion in a stream of oxygen and measurement of the resulting carbon dioxide by gravimetric, conductometric, or other means. If the sample is –100 mesh and the furnace temperature about 1125°C, iron drillings and cupric oxide are added [3]; –200 mesh samples usually burn readily at about 985°C without accelerators [4].

Free carbon in these carbides may be determined by several methods [4, 5]; the following procedure is typical. Place 1–5 g of –100 mesh sample in a platinum dish, add 10 ml hydrofluoric acid and 10 ml nitric acid, and evaporate to a low volume. Add more acid if necessary to bring about complete solution. Add 10–15 ml of a saturated boric acid solution, warm for a few minutes, and filter through asbestos on a Gooch crucible previously washed with hydrochloric acid and ignited. Wash the residue with hot water, hot 10% sodium hydroxide, hot water, hot 10% hydrochloric acid, and finally with hot water. Dry for 30 minutes at 105°C, and determine carbon by combustion in oxygen as usual.

A simple separation and determination of free carbon in materials containing refractory compounds has been published [6]. The sample is boiled with a mixture of 1:1 sulphuric acid and 5% potassium dichromate for 2–3 hours. The chromium (iii) produced is oxidized with excess standard potassium permanganate, and the latter back-titrated with standard ferrous ammonium sulphate.

In beryllium carbide, free carbon has been determined by refluxing with 1:1 sulphuric acid, filtering through asbestos in a crucible, and igniting at 900°C in a combustion tube. The combined carbon of the carbide is found by catalytic oxidation of the methane formed by an acid hydrolysis of the carbide [7]. Beryllium carbide liberates only methane by hydrolysis, and free carbon is not attacked by hydrolysis.

The total carbon in silicon carbide has been obtained by combustion in oxygen at 1350–1400°C; free carbon is measured by the same technique at a temperature of 750°C [8].

4. Organic and Inorganic Carbon

In shales and carbonates, the inorganic carbon can be removed by boiling with hydrochloric acid, after which the organic carbon is determined by a Leco induction furnace and volumetric carbon analyser. Total carbon can be determined in an identical sample by eliminating the hydrochloric acid treatment [9].

In another method, carbonate or inorganic carbon is determined by boiling the sample with phosphoric acid, aspirating the gas through standard barium hydroxide solution, and titrating the excess of the latter with standard hydrochloric acid, using thymolphthalein as indicator. For organic carbon, carbonate carbon is first removed, and then phosphoric acid plus chromium trioxide is used in place of the phosphoric acid above to oxidize the organic carbon [10].

Total and organic carbon in shale may be found by the following procedure. Weigh a sample into a small crucible for the Leco high-frequency combustion furnace. Add 1:4 hydrochloric acid to eliminate carbon dioxide of carbonates, and dry the sample at 100°C for 30 minutes to remove the excess hydrochloric acid. Add iron powder and tin metal as accelerators, and burn the sample in an oxygen stream at 1400°C to determine the carbon dioxide evolved from organic carbon. For total carbon, carry out the same procedure but omit the acid treatment [11].

REFERENCES

1. YOUNG, R. S., *Chemical Analysis in Extractive Metallurgy*, London, Charles Griffin, 1971.
2. YOUNG, R. S., SIMPSON, H. R., and BENFIELD, D. A., *Anal. Chim. Acta* **6**, 510–16 (1952).
3. FUREY, J. J., and CUNNINGHAM, T. R., *Anal. Chem.* **20**, 563–70 (1948).
4. TOUHEY, W. O., and REDMOND, J. C., *Anal. Chem.* **20**, 202–6 (1948).
5. FEICK, G., and GIUSTETTI, W., *Anal. Chem.* **36**, 2168–9 (1964).
6. MASHKOVICH, L. A., and KUTEINIKOV, A. F., *Khim. Svoistva Metody Anal. Tugoplavkikh Soedin.* **1969**, 18–26. *C.A.* **73**, 94329 (1970).

7. REED, S. A., FUNSTON, E. S., and BRIDGES, W. L., *Anal. Chim. Acta* **10**, 429–42 (1954).
8. SERRINI, G., and LEYENDECKER, W., *Met. Ital.* **64**, 129–37 (1972). *C.A.* **77**, 109038 (1972).
9. FOSCOLOS, A. E., and BAREFOOT, R. R., *Geol. Surv. Can., Pap.* **1970**, No. 70–1, 1–8.
10. BUSH, P. R., *Chem. Geol.* **6**, 59–62 (1970).
11. WIMBERLEY, J. W., *Anal. Chim. Acta* **48**, 419–23 (1969).

Chromium

Very few papers have appeared on phase analysis of chromium. Chromium nitrides in metallic chromium have been determined by the insolubility of Cr_2N and CrN in 10% perchloric acid [1].

In materials of the titanium industry, it has been found that, in a $CrCl_3$–$CrCl_2$ mixture, $CrCl_2$ only was dissolved in a solution of absolute alcohol and nitric acid [2].

REFERENCES

1. YANAGIHARA, T., MATANO, N., and FUKUDA, Y., *Nippon Kinzoku Gakkaishi* **27**, 152–6 (1963). *C.A.* **67**, 50124 (1967).
2. TIKHONOVA, A. P., and GUZ, L. Y., *Sb. Tr., Vses. Nauch.-Issled. Proekt. Inst. Titana* **1969**, 3, 294–9. *C.A.* **72**, 139302 (1970).

Cobalt

In the extractive metallurgy of cobalt, it is frequently necessary to determine metal, sulphide, or oxides when these forms occur together.

1. Metallic Cobalt

A. With Mercuric Chloride Solution

To 2 g of –200 mesh slag, add 50 ml of 7% mercuric chloride solution. Boil for 1 minute, filter through Whatman No. 42 paper, and wash thoroughly with hot water. Metallic cobalt displaces the mercury in mercuric chloride and passes into the filtrate as cobalt chloride, where it can be determined by any suitable procedure. Cobalt sulphide, cobalt silicate, and cobaltosic oxide, Co_3O_4 are virtually insoluble in the mercuric chloride solution. Cobaltic oxide, Co_2O_3, is soluble only to the extent of 0·1–0·3%; cobaltous oxide is slightly more soluble [1].

B. With Chlorine–Alcohol Solution

A differentiation of metal and sulphide from oxide in roasted and reduced ores and concentrates, based on leaching with a chlorine–alcohol solution, is sometimes employed in the base metal industries. For cobalt, however, in the presence of sulphide, this procedure has a limited value because the time required to leach all the sulphide also results in considerable solution of the higher oxides [2]. For example, when the same particle size and experimental techniques are used with a 0·25 g sample, one 10-minute leach will dissolve 99% of the metallic cobalt, about 80% of the cobalt sulphide, 0·2% of the cobaltous oxide, CoO, and 1–3% of the higher oxides Co_2O_3 and Co_3O_4. To attain 99% extraction of the sulphide, three 10-minute leaches are required; the solubility of the oxides then increases to 2–3% for CoO, 10–20% for Co_2O_3, and 8–11% for Co_3O_4.

In the absence of cobalt sulphide, the method separates metallic cobalt from its oxides; if the higher oxides of cobalt are not present, cobaltous oxide can be differentiated from the metal and sulphide as follows.

Weigh out a suitable quantity of –200 mesh sample, depending on the analytical procedure to be employed for cobalt; transfer to a dry 400-ml tall beaker. Add at least 10 times the sample weight of anhydrous methanol, stir, and place in a fume cupboard. Introduce a vigorous stream of gaseous chlorine from a cylinder into the dilute pulp for 10 minutes. Cobalt in metallic form, and most of the sulphide cobalt, will dissolve in the chlorine–alcohol solution, leaving the oxides virtually unattacked. Filter, and wash the residue with anhydrous methanol; carefully evaporate the filtrate to dryness and acidify with nitric acid. Make up to a definite volume and determine cobalt in a suitable aliquot.

If cobalt sulphide is present, a second or third 10-minute leach may be required to dissolve all of this compound; in this case the procedure is suitable only when cobaltous oxide is the sole form of cobalt oxide present.

C. With Bromine–Methanol

Metallic cobalt can be differentiated from cobalt oxides in reduced ores and oxides by the solubility of the metallic form and the insolubility of oxide in bromine–methanol solution. Cobalt sulphide, if present, is also soluble, and will report with the metallic cobalt. To 0·5 g of —100 mesh dry sample in a small Erlenmeyer flask add 50 ml of anhydrous methanol containing 5% bromine. Boil under a reflux condenser for 15 minutes, cool, filter, and wash with about 50 ml of methanol. To the filtrate add 20 ml of 1:1 hydrochloric acid, 5 ml of 75% hydroxylamine hydrochloride, and mix. Allow to stand until reduction of bromine is complete. Add 3% hydrogen peroxide, boil for 5 minutes to decompose excess hydrogen peroxide, cool, dilute to 200 ml, and determine cobalt by atomic absorption. Solution of oxide phases is not observed if free and combined water is removed from the sample [3].

2. Oxide Cobalt in Ores and Concentrates

In the extractive metallurgy of cobalt, it is sometimes necessary to differentiate the oxidized from the sulphide state in natural forms of the element. For instance, it may be required to obtain a measure of the ratio of oxidized minerals, such as asbolite, heterogenite, sphaerocobaltite, and stainierite, to the cobalt sulphides, carrollite, linnaeite, and siegenite. The following procedure, based on the selective solvent action of dilute sulphuric or hydrochloric acids on oxidized cobalt minerals in the presence of the reducing agent sulphurous acid, is applicable to ores and concentrator products [4].

Depending on the quantity of oxide cobalt present in the material and the analytical method used, select a weight of —200 mesh sample that will enable an accurate determination of cobalt to be made on the leached portion. The ratio of leach solution to sample may vary considerably, depending on the quantity of oxide cobalt present and the other acid-consuming constituents of the ore. In general, add 15–25 ml of 10% by volume hydrochloric acid saturated with sulphur dioxide, or 5% by volume sulphuric acid saturated with sulphur dioxide, to each gram of material in a stoppered Erlenmeyer flask or covered beaker. When the initial effervescence and attack has subsided, add 0·1–0·3 ml hydrofluoric acid per gram of sample. For large samples, these quantities of hydrofluoric acid can be kept to a minimum.

Shake the flask or beaker for about 10 seconds every 10 minutes for 1 hour; allow to stand for another hour, and agitate again at intervals for 10 minutes for the third hour. Filter the sample through Whatman No. 40 paper, with pulp, and wash thoroughly with hot water. Add 10 ml 1:1 sulphuric acid to the filtrate, boil out sulphur dioxide, and evaporate to strong fumes of sulphuric acid. Proceed with the determination of cobalt by any appropriate method. The leaching period will depend on the composition and mineralogy of the ore, and for many samples the time can be reduced substantially.

3. Differentiation of Cobalt Oxides

The proportions of cobaltous oxide and the higher oxides Co_2O_3 and Co_3O_4 in a mixture resulting from refining operations can be determined in the following manner [5]. Place 1 g of –200 mesh sample in a 250-ml Erlenmeyer flask with 20 ml of water and shake gently until all particles are completely wetted. Add 30 ml glacial acetic acid and attach a reflux condenser to the flask. Boil quietly for 1 hour. Pour the contents of the flask on to a tared sintered glass crucible and wash well with hot water. Dry to constant weight in an oven at 105°C. If the sample is impure, determine cobalt in the residue.

Under these conditions, cobaltous oxide is soluble in the dilute acetic acid, whereas the higher oxides are not. This treatment leaves cobalt sulphide practically intact [6], and consequently affords a procedure for separating sulphide from cobaltous oxide. When metallic cobalt is present, about 20% dissolves; its absence is essential for this differentiation.

REFERENCES

1. Young, R. S., *Chemical Analysis in Extractive Metallurgy*, London, Charles Griffin, 1971.

2. Young, R. S., *Chemist-Analyst* **49**, 46 (1960).
3. Kinson, K., Dickeson, J. E., and Belcher, C. B., *Anal. Chim. Acta* **41**, 107–12 (1968).
4. Young, R. S., Hall, A. J., and Talbot, H. L. *Am. Inst. Mining Met. Eng. Tech. Publ.* 2050 (1946).
5. Young, R. S., *The Analytical Chemistry of Cobalt*, Oxford, Pergamon, 1966.
6. Young, R. S., and Simpson, H. R., *Metallurgia* **45**, 51 (1952).

Copper

1. Forms of Copper in Mining and Extractive Metallurgy

Many copper minerals contribute to the economic importance of this metal, and diverse processes are employed in its extractive metallurgy. It is not surprising, therefore, that a number of publications have appeared on the phase analysis of both the natural forms of copper and those produced in the winning of the metal from its ores.

A. "Oxide" Copper, or "Acid-Soluble" Copper

In many ore deposits throughout the world, copper occurs in both oxidized and sulphide forms, and because their recovery methods differ, it is necessary to distinguish these forms. This can be done chemically with a reasonable degree of accuracy, because dilute sulphuric acid saturated with sulphur dioxide dissolves oxidized forms of copper but does not attack sulphides. An exception is the mineral cuprite, Cu_2O, which liberates only approximately half of its copper in the dilute acid leaching solution. On the rare occasions when appreciable quantities of cuprite are present, therefore, this method gives a low figure for oxide copper [1].

Weigh 0·5–2 g of –100 mesh sample into a 250-ml Pyrex "copper" flask, or Erlenmeyer, and add about 30 ml of 5% by volume sulphuric acid saturated with sulphur dioxide. Cover, and let stand for one hour, swirling occasionally. Filter into a 250-ml beaker and wash 6 times with hot water. The filtrate contains all the copper from malachite, azurite, tenorite, chrysocolla, and half the cuprite. Sulphides and native copper are not attacked, and remain in the precipitate. Boil off excess sulphur dioxide from the filtrate, and determine copper by any appropriate procedure. This is the "oxide", or "acid-soluble", copper frequently quoted in the industry. Determination of sulphide and metallic copper, plus half the cuprite, can be carried out on the residue from the oxide separation; alternatively, it may be obtained by subtracting the oxide value from that for total copper, determined on a separate sample.

B. Chrysocolla

The determination of chrysocolla, a copper silicate, in copper ores is of importance to a few producers because it is not amenable to normal flotation processes. Transfer a 1–5 g portion of $-100 + 200$ mesh ore to a 150-ml separatory funnel, and add 20–40 ml of tetrabromoethane. Stir thoroughly with a glass rod, and wash down any particles adhering to the walls of the funnel and to the stirring rod with a further 20–40 ml tetrabromoethane. If the mineral particles do not appear to be wetted readily, shake the funnel vigorously, and rinse the walls with additional tetrabromoethane.

Allow the material to remain in the liquid for an hour, occasionally tapping the funnel lightly to release any particles held mechanically in the upper or lower layers. Chrysocolla, together with other light minerals having a specific gravity below 2·96, will be on top of the tetrabromoethane; all other economically important copper minerals will sink to the bottom of the funnel. Draw off the bottom layer into a beaker, together with most of the tetrabromoethane; the latter can be recovered for re-use. Into another beaker transfer the float portion, washing any particles from this top layer, which may be adhering to the funnel, into the beaker.

Filter through a Whatman No. 40 paper, and wash thoroughly to remove tetrabromoethane. Dissolve the paper and precipitate in nitric and sulphuric acids, and evaporate to strong fumes of the latter. Determine copper by any appropriate method; copper $\times$ 2·76 = chrysocolla [2].

C. Metallic Copper

In the segregation process of the copper mining industry, where ores refractory towards conventional flotation are converted to segregated metallic copper, the latter is determined by the displacement of silver from a silver nitrate solution and measurement of dissolved copper by atomic absorption [3, 4]. The sample is shaken for 5 minutes with 100 ml of 16 g/litre silver nitrate solution, filtered into a litre flask and made to volume.

D. Various Copper Minerals

It has been reported that when a -325 mesh sample of copper ore is treated in the conventional manner to remove "oxide copper", and the residue is placed in a flask with 11 g thiourea, 50 ml N hydrochloric acid, and shaken for 3 hours, chalcopyrite remains insoluble whereas other copper sulphides are dissolved [5].

Another investigation with thiourea indicated that treatment with 6% thiourea solution dissolved oxide copper, tetrahedrite, and secondary sulphides, while primary copper sulphides are left as a residue [6].

Chalcocite has been separated from bornite and chalcopyrite by stirring the −150 mesh sample for 30 minutes with a 5% solution of unithiol containing 4–4·5 ml of 25% ammonium hydroxide with each 46 ml of unithiol. About 95–96% of the chalcocite is extracted, with no significant solution of bornite and chalcopyrite [7].

The separation of cuprite from tenorite and metallic copper has been effected by 1% aqueous unithiol, in which cuprite dissolves [8]. Tenorite is separated from metallic copper by the solubility of the former in 1% aqueous unithiol containing 5 g ammonium chloride per 100 ml.

When an ore sample containing chrysocolla, tenorite, and copper sulphides was agitated for 45 minutes in a 1% solution of unithiol containing 5 volume % hydrochloric acid, chrysocolla and tenorite dissolved completely, whereas less than 5% of the copper sulphides went into solution [9].

In another investigation using unithiol, when 0·5 g ore was leached for an hour with 5 ml of 1% water-glass + 25 ml of 1% unithiol, malachite and chrysocolla dissolve, leaving the copper sulphides in the residue [10].

It has been reported that in copper tailings the free uncombined copper oxides can be dissolved by leaching with dilute sulphuric acid containing sodium sulphide and an organic inhibitor ChM which is a derivative of quinoline, pyridine, naphthalene, and a porphyrine-like compound; combined oxidized copper minerals dissolve in dilute sulphuric acid containing sodium sulphite and ammonium bifluoride, leaving sulphides in the residue [11].

A different approach to phase analysis of copper ores has been published [12]. A water leach removes chalcanthite, then agitation for 1 hour at room temperature with 5% sulphuric acid containing sodium sulphite dissolves malachite and azurite; agitation of the residue for 2 hours with 5% sodium cyanide leaches covellite, leaving chalcopyrite in the final residue. An observation in an earlier paper, that the presence of manganese dioxide increases the solubility of chalcocite and covellite in potassium cyanide solution, may have application here [13].

In another procedure, boiling the −200 mesh sample for 2 hours in a nitrogen atmosphere with alkaline 10% sodium potassium tartrate dissolves malachite. The residue is washed with 1% sodium hydroxide and boiled for another 2 hours in an oxygen atmosphere to dissolve cuprite. Stirring the residue for 3 hours with 5% sulphuric acid and 2% ferric sulphate dissolves 50% of the chalcocite. The third residue is then boiled for 10 minutes with 5 ml of acetic acid and

100 ml of saturated silver sulphate, and stirred for 4 hours to dissolve the rest of the chalcocite and the bornite. The final residue is chalcopyrite [14].

A separation of chalcocite from bornite has been described [15]. When 1 g of the sample is agitated with 50 ml of 1% silver nitrate, the copper from chalcocite goes into solution as copper nitrate, leaving bornite in the residue.

A scheme for differentiating the copper compounds formed, when copper–zinc sulphides high in iron are roasted, has been published [16]. A water leach dissolves copper sulphate, and when the residue is agitated with 15% ammonium chloride solution, cuprite, metallic copper and part of the basic copper sulphate go into solution. A leach of the second residue for 1 hour with 5% acetic acid brings tenorite into solution, leaving copper sulphides in the final residue.

A paper on the phase analysis of chlorination roasting products for copper, zinc, and lead compounds has been published [17], but no details are given in the abstract.

2. Differentiation of Copper and its Oxides in Powders

Interest in the measurement of metallic copper, cupric oxide, and cuprous oxide, originally confined to powders for marine pigments and catalysts, was greatly intensified with the development of powder metallurgy in metal fabricating, and of pressure hydrometallurgical processes in extractive metallurgy wherein metals are frequently recovered in powder form. A number of methods for such a differentiation have been published.

Cuprous oxide in a mixture with cupric oxide has been determined by digesting 0·5 g with 50 ml of 6N sulphuric acid and 50 ml of 0·1N potassium permanganate for 30 minutes at 30–40°C. Add 15 ml of 0·1N oxalic acid, and titrate the excess at 70–90°C with standard potassium permanganate [18]. Cupric oxide was determined by the relation: total copper − cuprous copper = cupric copper.

Metallic copper in a mixture with oxides has been found by converting it to cupric sulphide by means of a solution of sulphur in carbon bisulphide, and separating this from the oxides by dissolving the latter in a mixture of potassium chloride and hydrochloric acid [19].

In another procedure, cuprous oxide is dissolved selectively in 3N ammonium chloride at 95°C under a carbon dioxide atmosphere. After filtration, the residue is digested with 6N sulphuric acid at 70°C to dissolve cupric oxide; metallic copper remains in the final residue [20, 21].

In a modification of the method above, treatment with 2% Trilon, or disodium ethylenediamine tetraacetate, under a nitrogen atmosphere, and vacuum filtration, brings copper salts into solution, leaving oxides in the residue; treatment of the latter three times with boiling 5% ammonium chloride with nitrogen agitation extracts the cuprous oxide, and 2% stannous chloride in hydrochloric acid with nitrogen agitation extracts cupric oxide from the second residue [22].

Metallic copper in oxides has been determined by adding an excess of iodine in 4M hydrochloric acid, and titrating the excess iodine with sodium thiosulphate [23].

It has also been found that cuprous ion can be determined, in a mixture with cupric, by complexing the latter with EDTA and following the usual iodometric method [24].

In a sulphuric acid medium, cuprous copper in the presence of cupric can be determined by an indirect potentiometric titration at 55°C, or an amperometric titration, with potassium permanganate solution [25].

Cuprous oxide in the presence of cupric is dissolved selectively by stirring a 0·5-g sample in a carbon dioxide atmosphere with 9 ml of 1:9 hydrochloric acid containing 3 ml of 3% hydrazine sulphate and 3 ml of 10% potassium chloride [26].

In a variation of this method, cuprous oxide in a mixture with metallic copper is dissolved under carbon dioxide by a solution containing dilute hydrochloric acid, hydrazine sulphate, and potassium chloride. Still under carbon dioxide, dissolve the residue of copper and cupric oxide in stronger hydrochloric acid and ferric chloride and determine the copper by titrating the ferrous ion formed. In another sample, determine the total reducing power of copper + cuprous oxide by dissolving in hydrochloric acid + ferric chloride and titrating the ferrous ion formed [27].

In a mixture of metallic copper and oxides, copper metal + cuprous oxide is obtained by dissolving the sample in sulphuric and phosphoric acids, adding excess potassium dichromate, and back-titrating with ferrous sulphate. In another sample, metallic copper is selectively determined by neutral 0·1N silver nitrate solution with addition of about 15% ethanol to suppress dissolution of copper oxides [28].

A reliable procedure for determining metallic copper, cuprous oxide, and cupric oxide in a mixture of these forms is outlined below [29, 30, 31]. Transfer a weighed portion to a 250-ml Phillips conical beaker with lip. Add 100 ml of an extraction solution containing 40 ml of hydrochloric acid and 40 g of stannous chloride dihydrate per litre of ethanol. Add lumps of dry ice to keep the temperature near 0°C, and swirl the flask for 5 minutes. Cuprous oxide is dissolved; any cupric oxide which dissolves is reduced by stannous chloride to the cuprous state without any attack on metallic copper. Filter the residue of copper + cupric oxide on asbestos, and wash with ethanol. Dissolve this residue by warming on a steam bath in 25 ml of a solution containing 75 g ferric chloride hexahydrate, 150 ml hydrochloric acid, 400 ml of water, and 5 ml of 30% hydrogen peroxide boiled to remove excess of the latter. Maintain an atmosphere of carbon dioxide above the solution by the addition of dry ice. Add 50 ml of water, 3 drops of *o*-phenanthroline, and titrate the ferrous iron with 0·1N ceric sulphate to the change from orange to pale green. The metallic copper reduces the ferric salt to ferrous; cupric oxide is left in the residue.

Metallic copper + cuprous oxide can be determined by digesting an original sample directly with 25 ml of the ferric chloride/hydrochloric acid solution for 15 minutes in a carbon dioxide atmosphere. The total copper in the sample may be found by any appropriate method. The total copper minus the sum of metallic copper + cuprous oxide gives cupric oxide; total copper minus the sum of metallic copper + cupric oxide gives cuprous oxide.

REFERENCES

1. Young, R. S., and Graham, D. G. M., *Ind. Eng. Chem. Anal. Ed.* **14**, 787–8 (1942).
2. Young, R. S., and Simpson, H. R., *Mining Mag.* **84**, 137–9 (1951).
3. Mackay, K. E., and Gibson, N., *Trans. Institution Mining Met.* **77**, C 19–31 (1968).
4. Smith, G. A., and MacLeod, D. S., *Trans. Institution Mining Met.* **77**, C 229–31 (1968).
5. Dolaberidze, L. D., and Politova, Y. V., *Zavod. Lab.* **32**, 779–81 (1966). *C.A.* **65**, 16049 (1966).
6. Timerbulatova, M. I., and Khristoforov, B. S., *Metody Izuch. Veshchestv. Sostava Ikh Primen.* **1969**, No. 2, 45–50. *C.A.* **71**, 119293 (1969).
7. Songina, O. A., Ospanov, K. K., and Rozhdestevnskaya, Z. B., *Zavod. Lab.* **32**, 782–3 (1966). *C.A.* **66**, 34604 (1967).
8. Ospanov, K. K., and Songina, O. A., *Zavod. Lab.* **34**, 159 (1968). *C.A.* **69**, 15764 (1968).
9. Ospanov, K. K., and Tember, N. I., *Zavod. Lab.* **34**, 662–5 (1968). *C.A.* **69**, 92600 (1968).
10. Pashevkina, O. N., and Gurkina, T. V., *Zavod. Lab.* **37**, 891–5 (1971). *C.A.* **75**, 14722 (1971).
11. Leont'eva, K. D., *Sb. Nauchn. Tr. Gos. Nauchn.-Issled. Inst. Tsvetn. Metal.* No. **18**, 118–26 (1961). *C.A.* **60**, 4778 (1964).
12. Ionescu, M., and Pavel, R., *Rev. minelor* **9**, No. 1, 39–44 (1958). *C.A.* **53**, 12092 (1959).
13. Fedorova, M. N., *Zavod. Lab.* **16**, 904–6 (1950). *C.A.* **49**, 12194 (1955).
14. Yen, J. Y., and Liu, Y. S., *Hua Hsueh Hsueh Pao* **25**, 346–52 (1959). *C.A.* **54**, 16270 (1960).

15. Solntsev, N. I., and Usova, L. V., *Sb. Tr. Gos. Nauchn.-Issled. Inst. Tsvetn. Metal.* No. **19**, 756–72 (1962). *C.A.* **60**, 3476 (1964).
16. Stroiteleva, G. P., and Suvorova, M. V., *Met. i Khim. Prom. Kazakhstana, Nauchn.-Tekhn. Sb.* **1961**, No. 4, 83–7. *C.A.* **58**, 3886 (1963).
17. Filippova, N. A., Martynova, L. A., Selezneva, M. N., and Stepareva, V. N., *Sb. Nauch. Tr. Nauch.-Issled. Inst. Tsvet. Metal.* **1971**, 34, 179–82. *C.A.* **77**, 172271 (1972).
18. Neubert, H., *Chemist-Analyst* **27**, 15 (1938).
19. Ubaldini, I., and Guerrieri, F., *Ann. Chim. Applicata* **38**, 695–701 (1948). *C.A.* **43**, 8969 (1949).
20. Crisan, I., *Bul. Inst. Politeh. Bucuresti* **21**, 115–20 (1959). *C.A.* **56**, 2889 (1962).
21. Podchainova, V. N., *Zhur. Anal. Khim.* **7**, 305–11 (1952). *C.A.* **47**, 2631 (1953).
22. Filippova, N. A., *Analiz Rud Tsvetnykh Metal. i Produktov ikh Pererabotki, Sbornik Nauch. Trudov* **1958**, No. 14, 169–78. *C.A.* **53**, 13878 (1959).
23. Rychcik, W., and Basinska, H., *Stud. Soc. Sci. Torun., Sect. B.* **1972**, 8, 1–8. *C.A.* **77**, 96546 (1972).
24. Subramanian, R., Venkatachalam, R., Chakrapini, S., and Shenoi, A. B., *J. Inst. Chem.* **43**, 201–7 (1971). *C.A.* **76**, 67679 (1972).
25. Makarov, G. V., *Tr. Khim.-Met. Inst. Akad. Nauk Kaz. SSR* **1970**, No. 16, 40–7. *C.A.* **76**, 41661 (1972).
26. Lobanov, F. I., Shibrya, G. G., and Dadugina, N. G., *Novye Metody Khim. Anal. Mater.* **1971**, No. 2, 9–11. *C.A.* **77**, 121748 (1972).
27. Baker, I., and Gibbs, R. S., *Ind. Eng. Chem. Anal.* Ed. **15**, 505–8 (1943).
28. Seeger, S. B., and Escibar, W. A., *Rev. Real Acad. Cienc. Exactas, Fis. Natur. Madrid* **1968**, 62, 205–30. *C.A.* **70**, 86192 (1969).
29. Baker, I., and Gibbs, R. S., *Ind. Eng. Chem. Anal.* Ed. **18**, 124–7 (1946).
30. Lavrukhina, A. K., *Zhur. Anal. Khim.* **1**, 73–9 (1946). *C.A.* **43**, 2543 (1949).
31. Young, R. S., *Chemical Analysis in Extractive Metallurgy*, London, Charles Griffin, 1971.

Germanium

Several publications have appeared on the phase analysis of germanium compounds.

A procedure for separating the forms of germanium in metallurgical flue dusts has been described [1]. To a 0·25–1 g sample is added 100 ml of 6·5M ammonium hydroxide + 0·4 g ammonium chloride, and the solution is gently agitated at room temperature for 30 minutes. The suspension is filtered; the solution contains germanium from GeO and GeS_2. The residue is repeatedly mixed with new portions of the same solution, filtered, and then boiled with 100 ml of 0·1M Complexon 111 solution for 10 minutes. After filtering, the solution contains germanium derived from germanates. The residue is heated with 50 ml of 3% hydrogen peroxide at 70–80°C for 1 hour, filtered, and washed. This solution contains germanium from GeS.

The determination of metallic germanium in rare earth germanates has been outlined [2]. For samples which require the use of hydrofluoric and sulphuric acid to effect decomposition, place a 0·1-g sample in a platinum dish with 2 ml 0·1N ammonium vanadate solution, 2–3 ml 6N sulphuric acid, 6 ml water, and 4 ml hydrofluoric acid. Heat until decomposition is complete, cool, transfer to a titrating flask with 30 ml water or saturated boric acid solution and 25 ml of 1:2 sulphuric acid. Add 5 ml of 1:1 phosphoric acid, 2 drops sodium diphenylamine sulphonate indicator solution, and titrate with 0·05N Mohr's salt to the disappearance of the lilac colour.

If the material can be decomposed by hydrochloric acid, place the sample in a beaker with 1 g sodium carbonate, 1 ml of 0·01N Mohr's salt, 5 ml of 0·06N ferric chloride solution, and 40 ml of hydrochloric acid. Boil gently until the sample is decomposed, cool, add 10 ml of 1:1 phosphoric acid, 2 drops of sodium diphenylamine sulphonate indicator, and titrate the ferrous iron with 0·01N potassium dichromate solution.

REFERENCES

1. Filippova, N. A., and Korosteleva, V. A., *Sb. Nauch. Tr., Gos. Nauch.-Issled. Inst. Tsvet. Metal.* **1968**, No. 28, 61–71. *C.A.* **70**, 16801 (1969).
2. Piryutko, M. M., and Kostyreva, T. G., *Zavod. Lab.* **38**, 1313–14 (1972). *C.A.* **78**, 92073 (1973).

Indium

A phase analysis of indium compounds in flue dusts has been published [1]. The dusts contained, in percentages, Pb 27–50, Zn 5–8, Cu 0·1–0·3, As 0·8–7, Sb, Se, and Te 0·02–0·12.

Agitate the sample with water for a few minutes at room temperature, filter, and wash; the filtrate contains indium sulphate, $In_2(SO_4)_3$. Treat the residue with 0·4M bromine in methanol to extract the sesqui-sulphide In_2S_3, and filter. The final residue contains indium in the form of the sesqui-oxide In_2O_3, which will go into solution when the residue is boiled for 30 minutes with 3N hydrochloric acid.

REFERENCE

1. Terzeman, L. N., *Zavod. Lab.* **38**, 1439–40 (1972). *C.A.* **78**, 79248 (1973).

Iron

The number of publications on the phase analysis of iron greatly exceeds that for any other element. There are several reasons for this. From the various forms of oxidized and sulphide iron which occur in many parts of the world in enormous deposits, man produces metallic iron; Nature slowly transforms the metal back to ferric oxide. The importance of a knowledge of the forms of iron to the geologist, mineralogist, and petrographer has resulted in the common practice of reporting ferrous and ferric iron in rock analyses for nearly a century. As befits the world's most important metal, iron and its compounds have been extensively investigated by primary producers, fabricators, and consumers. Iron is a common impurity in ores, and methods for its elimination in recovery processes require a knowledge of its phases. Finally, developments in powder metallurgy have given a substantial impetus to investigations into the phases of iron.

1. Metallic Iron

Metallic iron is rarely found in nature and therefore seldom encountered in the analysis of rocks, minerals, and ores. In the production of iron, or preparation of iron powder, by reduction of oxides, the determination of metallic iron is a frequent and important analysis. Many methods have been proposed for the metallic phase of iron; the principal ones are outlined below.

A. Mercuric Chloride

The oldest procedure, dating from 1880, uses a solution of mercuric chloride to dissolve the metal phase; the mercurous chloride formed is removed and the ferrous iron titrated. It is essential to use a finely pulverized sample, preferably −200 mesh.

The basic procedure, advocated by a number of investigators [1, 2, 3, 4, 5, 6, 7] may be outlined as follows. To 0·5–2 g of −200 mesh material, depending on the anticipated metallic iron content, add 50–100 ml of 5% mercuric chloride solution. Heat for 10–20 minutes at about 95°C, cool rapidly, filter through Whatman No. 42 paper, and wash thoroughly with hot water. Iron oxides remain in the residue. The iron in the filtrate can be determined by titration with potassium permanganate or potassium dichromate in the conventional manner.

In a refinement of the method, the reaction between the sample and the mercuric chloride solution can be carried out in an atmosphere of carbon dioxide. The flask is closed with a Bunsen valve, cooled, filled with water, mixed, allowed to settle, and filtered into a vessel containing carbon dioxide. An aliquot is acidified with sulphuric acid, and the ferrous chloride titrated with standard potassium permanganate. The amount of chloride present in the filtrate is so small that addition of manganous salt is usually unnecessary.

Variations of the above method have been published. In one, metallic iron is extracted with mercuric chloride in ethanol; in the presence of oxidants and calcium oxide, mercuric chloride and sodium salicylate in methanol are used [8].

Metallic iron in reduced manganese agglomerates has been determined by heating the sample with 0·6 g mercuric oxide and 40 ml of 5% mercuric chloride in acetone for 15 minutes at 56°C, adding 6–8 drops acetylacetone, boiling 5–7 minutes and diluting to 100 ml with the mercuric chloride/acetone solution. Iron is finally measured colorimetrically on a filtered aliquot against the blank of mercuric chloride/acetone solution [9].

In another modification, metallic iron is dissolved in an alcoholic mercuric chloride solution made by dissolving 50 g of mercuric chloride, 40 g lithium chloride, 20 g ammonium chloride, and 15 g of sodium salicylate in anhydrous ethanol, and diluting to 1 litre. The dissolved iron is oxidized with ammonium persulphate and titrated with a titanous salt solution [10].

Another approach has been used for metallic iron in meteorites, by dissolving iron, nickel, and cobalt in a mercuric chloride/ammonium chloride solution, making the extract 10M in hydrochloric acid and separating the three elements by elution from an anion exchange column [11].

B. Copper Salt

The displacement of copper from cupric sulphate or another copper salt by metallic iron forms the basis for another popular and satisfactory method for metallic iron.

The procedure outlined below is representative of many recorded in the literature [4, 12, 13, 14, 15, 16, 17]. Weigh a –200 mesh sample into a 400-ml beaker, and add 20–25 ml of 10% copper sulphate solution. The latter is usually neutral but may contain 0·075% sulphuric acid. Dilute with 25–30 ml hot water, cover the beaker and boil for

15 minutes, maintaining the volume of approximately 50 ml by addition of water. Filter, wash with hot water, dilute to about 100 ml, add 14 ml 1:1 sulphuric acid and a strip of aluminium. Boil until all copper has precipitated, cool, filter, wash with cold water, dilute to about 250 ml with cold water, and titrate with standard potassium permanganate solution. This figure for iron represents the metallic iron in the sample. Oxides do not interfere, but the presence of ferrous sulphide causes high results.

Many variations of the copper–salt procedure have been published. An atmosphere of carbon dioxide in the reaction vessel is frequently used [15, 18, 19, 20]. Potassium dichromate is often employed as titrant in place of permanganate [13, 21, 22]; in this case sodium diphenylamime sulphonate is used as indicator.

For slags, it has been suggested that the sample be boiled for 40 minutes in copper sulphate solution under reflux [21].

An improvement in the determination when iron sulphide is present is effected by the use of a mixed cupric chloride/potassium chloride solvent for metallic iron [18, 19]. Place the –100 mesh sample in a 125-ml Erlenmeyer flask from which the air has been swept out by carbon dioxide. Pour 30 ml of a solution containing 53·35 g $CuCl_2.2H_2O$ and 46·65 g KCl per litre into a beaker, add 5 drops of glacial acetic acid, and saturate with carbon dioxide by bubbling the gas through for several minutes. Add this to the Erlenmeyer flask, and agitate for 30 minutes with a steady stream of carbon dioxide. Swirl occasionally to prevent caking. Using a platinum cone and a coarse paper, filter the solution under light suction into a 300-ml Erlenmeyer flask. Rinse the original flask and the filter with about 100 ml of water. Add 1 g of pure aluminium shavings and 6 ml of 1:1 hydrochloric acid. Boil gently under a carbon dioxide atmosphere until the solution becomes colourless, then cool under the same condition in an ice bath. Filter with gentle suction through a fine paper into a 500-ml Erlenmeyer flask, and wash the residue with cold water. Add 10 ml of Zimmerman–Reinhardt reagent and titrate with standard potassium permanganate solution to a faint pink. The presence of FeS does not affect this determination, but Fe_3C does.

In another publication, a combination of silver sulphate and copper sulphate is recommended for metallic iron in oxidized iron powders [23]. To a 0·5-g sample add 40 ml of 0·5% silver sulphate, shake for 35 minutes, add 30 ml of 30% copper sulphate, and shake for 30 minutes. Filter and titrate with potassium dichromate, using sodium diphenylamine sulphonate indicator. The silver sulphate is stated to penetrate more efficiently through the oxide layer.

A different approach is taken in another publication [24]. The sample of ore or slag is shaken for 1 hour with a copper sulphate solution containing 1 g of mercury. After filtration, the copper equivalent of the metallic iron is determined in the precipitate, and the percentage of iron equals

$$\% \text{ of copper} \times \frac{\text{at. wt. of iron}}{\text{at. wt. of copper}}$$

The mercury addition is stated to keep the surface of metallic iron free from deposited copper, and to act as a catalyst in the reaction.

It has been reported that metallic iron with natural pyrrhotite yielded correct results using the copper sulphate method, but high values were obtained for metallic iron in the presence of synthetic troilite due to solution of iron sulphide [20].

In another paper, it was found that metallic iron, siderite, and hydrosilicates are soluble in copper sulphate solution, but oxides and hydroxides of iron are insoluble [25].

After solution of metallic iron in neutral copper sulphate, and filtration, the ferrous iron can be titrated with standard ceric sulphate solution [15].

C. Bromine–Alcohol

The solubility of metallic iron, and insolubility of its oxides in a 2–5% bromine solution in anhydrous ethanol or methanol provides a good procedure for the phase analysis of this metal. Sulphides are likewise soluble and accompany metallic iron. A representative method is outlined below [26–38].

To 0·5–2 g of the −100 mesh dry sample in an Erlenmeyer flask add 50 ml of anhydrous methanol containing 5% bromine. Boil for 15 minutes under a reflux condenser, cool, filter, and wash thoroughly with anhydrous methanol. Metallic iron passes into the filtrate and can be determined by any appropriate method. The filtrate may be evaporated to dryness, treated with nitric acid to destroy organic matter, and the iron finally dissolved in hydrochloric acid and titrated, after reduction, by potassium dichromate or permanganate.

In this, as in all phase analyses involving the use of a bromine–alcohol solution, great care should be exercised when mixing these reagents. A rapid exothermic reaction can result in the violent ejection of the mixture from the vessel. It is usually good practice to place the beaker containing methanol in a bath of cold running water for a short period before adding the bromine, with constant stirring. The mixture may then be allowed to return to room temperature.

Several variations in the procedure have been published. Ammonium peroxydisulphate can be added to the filtrate containing metallic iron, the *p*H adjusted to 2, and the iron titrated with EDTA, using salicylic acid as indicator [29].

In another method, 20 ml of 1:1 hydrochloric acid and 5 ml of

75% hydroxylamine hydrochloride are added to the filtrate containing metallic iron. The solution is mixed and allowed to stand until reduction of bromine is complete. Hydrogen peroxide is slowly added until the straw colour of the chloroferrate complex is fully developed, the solution is boiled for 5 minutes to decompose excess hydrogen peroxide, cooled, diluted to an appropriate volume, and the iron determined by atomic absorption [30].

It has been recommended to prepare bromine–methanol solution by adding bromine to anhydrous methanol which contains 10% formic acid, and shaking for a period just before use [31].

For metallic iron in electrocorundum, a 1-g sample is refluxed for 1 hour, with periodic shaking [32].

Iron in the filtrate from a bromine–methanol separation can be determined iodometrically [35].

One preparation of bromine–methanol solution calls for a mixture of 50 ml methanol, 2 ml bromine, and 10 ml of acetone, diluted to 100 ml with methanol [36].

D. Hydrogen Evolution

Several papers have appeared on the determination of metallic iron by measuring the volume of hydrogen evolved with hydrochloric acid. The materials investigated were sponge iron [39], minerals [40], and a mixture of metallic iron, ferrous sulphide, and ferric and ferrous oxides [41]. In the latter mixture, hydrogen sulphide was also evolved on acid treatment, and was separated from hydrogen by absorption in ammoniacal cadmium chloride solution.

E. Silver Thiocyanate

A few workers have reported the determination of metallic iron by its solubility in silver thiocyanate or silver ammonium thiocyanate. Iron displaces silver and is determined by titration of the ferrous ion formed.

Typically, metallic iron is dissolved at room temperature, in a solution containing, per litre, 120 g ammonium thiocyanate, 40 g silver thiocyanate, 40 g ammonium chloride, 8 g sodium acetate, and 8 ml acetic acid. After filtration, thiocyanate is removed by excess mercuric nitrate and ferrous iron is titrated with potassium dichromate [42].

The silver ammonium thiocyanate may also be dissolved in a water–dioxane solvent [43].

Iron in the filtrate may be determined photometrically with ascorbic acid and *o*-phenanthroline [33].

In an evaluation of methods for the determination of metallic iron in reduced iron ores, it has been reported that the silver thiocyanate method yields high results [38].

F. Ferric Chloride

The selective solvent action of a dilute ferric chloride solution on metallic iron, leaving iron oxides unattacked, has been utilized in phase analyses [44–50]. Usually, the sample is boiled for an hour in 6% ferric chloride solution, cooled, filtered, washed, and the ferrous iron in the filtrate titrated with potassium dichromate; this represents the metallic iron.

It has been reported that when this reaction is carried out in an atmosphere of carbon dioxide very little attack occurs on the oxides, carbides, silicides, sulphides, phosphides, and nitrides of iron [45].

The final determination of ferrous iron, representing the metallic iron, can be done potentiometrically with a bromate solution [46].

A leaching period of 20 hours at room temperature can be substituted for a one-hour boiling time [47].

It has been observed that the usual 6% ferric chloride solution also dissolves metallic aluminium, cobalt, copper, manganese, molybdenum, nickel, tungsten, and vanadium [48].

In the presence of magnetite, metallic iron, and iron carbonate, the latter two are dissolved in ferric chloride solution; the method cannot be used with ferruginous silicates [49].

For metallic iron in meteorites, a mixture of ferric chloride, ferric ammonium sulphate, and ammonium sulphate has been used as the solvent [50].

G. Miscellaneous Methods

Metallic iron in the presence of iron oxides has been determined by the action of lead chloride solution and titration of the resulting ferrous iron with potassium dichromate and diphenylamine [51].

Metallic iron in a sample with iron oxides has been dissolved by stirring at room temperature in a mixture of 10% potassium chlorate and 52% nitric acid [52, 53].

In the presence of iron hydroxide, oxyhydroxide, and oxide, metallic iron has been determined on the residue after dissolving the former three by agitation for 5–10 minutes in 3N hydrochloric acid [54].

It is possible to separate metallic iron from ferrous and ferric oxides by means of a laboratory magnetic separator. Several passes through the separator, to remove non-magnetic particles which are carried mechanically into the magnetic fraction, will generally yield a good differentiation; from the weight of fractions and the content of iron the metallic iron in the sample may be calculated. Such a separation fails, of course, in the presence of magnetite and of many pyrrhotites.

2. Ferrous Iron

The determination of ferrous iron in solution is, of course, a very simple one, involving merely the addition of standard potassium permanganate solution to the first permanent pink, or the addition of potassium dichromate in the presence of a diphenylamine indicator to the violet-blue colour. Most volumetric iron procedures are based on a decomposition to bring all iron into solution, the reduction of all the iron present to the ferrous state, followed by titration with a standard oxidant. In solutions derived from metal leaching, refining, or pickling operations, ferrous iron may accordingly be determined directly by adding to a measured volume 15 ml of 20% sulphuric acid, 2–3 drops of 0·2% sodium diphenylamine sulphonate indicator, and titrating with standard potassium dichromate to the appearance of the violet-blue endpoint.

When ferrous iron must be determined on a solid sample which requires acid decomposition or other means to solubilize the iron, however, the phase analysis becomes more complicated. It becomes even worse when, in addition to ferrous and ferric iron, the sample contains metallic iron, iron sulphides, etc.

For materials such as rocks, minerals, and ores, which have little or no metallic iron, there are two principal methods for determining ferrous iron. In one, decomposition is carried out by a boiling sulphuric–hydrofluoric acid mixture in a closed platinum crucible; the contents of the latter are transferred to a beaker of water containing sulphuric, phosphoric, and boric acids, and the ferrous iron is titrated with standard potassium permanganate or dichromate solution.

Details of this old-established method are given below [55, 56, 57]. Place 0·5 or 1 g of —100 mesh material in a large platinum crucible and moisten with a few drops of freshly boiled and cooled water. Add several drops of 1:1 sulphuric acid, cover, and allow to stand until all reaction has ceased. Add 10 ml of 1:1 sulphuric acid, cover with a close-fitting platinum lid, and heat rapidly until the contents are nearly boiling. Carefully move the lid to the side and quickly but cautiously add 5 ml hydrofluoric acid, cover again immediately, and boil gently for 10 minutes. In the meantime, to about 150 ml of freshly boiled and cooled water in a 400-ml beaker add 10 ml of 1:1 sulphuric acid, 5 ml of phosphoric acid, and 25 ml of a saturated solution of boric acid.

At the end of the 10-minute heating period, quickly immerse the covered crucible in the solution in the 400-ml beaker; remove and rinse the crucible and lid. Add 3 drops of 0·2% aqueous solution of sodium diphenylamine sulphonate indicator and titrate the ferrous iron with standard potassium dichromate solution to the violet-blue endpoint.

If preferred, standard potassium permanganate solution may be used as titrant in place of dichromate. In this case, of course, no indicator is required; the endpoint is the first permanent pink imparted by the permanganate itself.

There are several modifications of this well-known procedure. Sometimes a stream of carbon dioxide is introduced through a small opening in the crucible lid during the initial heating period; after the solution commences to boil the steam prevents ingress of air and the carbon dioxide flow is discontinued. Other workers use a crucible lid having a small opening to allow the escape of steam, but do not introduce carbon dioxide. Many analysts use a regular crucible lid, without any hole, but position the lid on the crucible so that a very slight opening may allow steam to escape but minimize the possibility of oxidation from the air.

Potentiometric titration, rather than the visual one, may be employed in another variation. The procedure up to the titration stage is the same as described above, except that the addition of 5 ml phosphoric acid to the solution in the 400-ml beaker is not necessary.

Place in the solution the platinum and saturated calomel reference electrodes of a potentiometric apparatus, and agitate the solution by means of a plastic-covered stirring bar and a magnetic stirrer. Titrate with standard potassium dichromate solution until a sharp increase in potential indicates that all the ferrous iron has been oxidized. The potential will gradually rise from about 400 mV initially to about 500 mV, after which the changes will become more pronounced; the abrupt rise at the end point usually occurs at 600–650 mV.

With care in the rate of heating, and in the addition of hydrofluoric acid, satisfactory results may be obtained by this procedure. In common with many phase analyses, the method often requires a little practice before acceptable reproducibility is achieved.

The determination of ferrous iron is, however, subject to errors inherent in the method. Metallic iron, whether a natural constituent of the sample or introduced from the pulverizing equipment, will be counted as ferrous iron and may also reduce some of the ferric iron present as well. It is impossible to accurately determine ferrous iron if sulphide minerals are present in the sample. Carbonaceous material, apart from graphite, will seriously interfere in the titration. Trivalent vanadium will likewise give a high value for ferrous iron. Another source of error is the presence of refractory iron-bearing minerals which are not decomposed in boiling sulphuric–hydrofluoric acid.

In the other method, ammonium vanadate solution and hydrofluoric acid are added to the sample in a plastic container, the sample is allowed to stand overnight, and transferred to a beaker of water containing sulphuric, phosphoric, and boric acids; the excess of pentavalent vanadium is titrated with standard ferrous ammonium sulphate, using sodium diphenylamine sulphonate as indicator. Alternatively, an excess of ferrous ammonium sulphate may be added, and the excess finally titrated with standard potassium dichromate.

A typical outline of the cold acid decomposition/ammonium vanadate method is given below [56, 58, 59]. Transfer 0·2 g of —100 mesh material to a small plastic flask, and add 5 ml of 0·1N ammonium vanadate solution; the latter is made by dissolving 10 g of ammonium metavanadate in 110 ml of 1:1 sulphuric acid and diluting to 1 litre. Add 10 ml of hydrofluoric acid, stopper the flask loosely with a plastic thimble, and allow to stand overnight. Add 10 ml of a sulphuric–phosphoric acid solution containing 400 ml of sulphuric acid and 200 ml of phosphoric acid per litre.

Pour the contents of the flask into a 400-ml beaker containing 100 ml of saturated boric acid solution. Rinse the flask with another 100 ml of boric acid solution, adding this to the 400-ml beaker. Add 10 ml of standard ferrous ammonium sulphate solution, 0·05N being convenient, and 3 drops of 0·2% sodium diphenylamine sulphonate indicator. Titrate with 0·05N potassium dichromate solution to a grey endpoint, using a small burette. Alternatively, the titration can be carried out by simply titrating the excess vanadate with standard ferrous ammonium sulphate, using sodium diphenylamine sulphonate as indicator.

As indicated earlier, metallic iron must be removed from a sample before a determination of ferrous iron can be carried out. In many cases, the removal and determination of metallic iron by one of the methods described earlier is followed by an analysis of the residue for ferrous iron. Examples of these will now be discussed.

When metallic iron is dissolved in mercuric chloride solution, leaving iron oxides in the residue, mercury can be removed from the latter by means of an alcoholic solution of iodine [7, 10]. The oxides are dissolved in sulphuric acid under an atmosphere of carbon dioxide, and ferrous iron determined by titration with standard potassium permanganate solution. Alternatively, under the same inert atmosphere the oxides can be dissolved by hydrochloric acid, and the ferrous iron determined by potassium dichromate, using sodium diphenylamine sulphonate as indicator.

If copper sulphate is used to determine metallic iron, the oxides are filtered off in the residue. After solution of oxides in sulphuric acid the ferrous iron may be titrated with potassium permanganate or potassium dichromate [16, 22]; decomposition may also be effected by hydrochloric acid with final titration by dichromate, or ceric sulphate using *o*-phenanthroline as indicator [13, 15].

When a bromine–alcohol solution is employed to dissolve metallic iron, the determination of ferrous iron in the residue may be carried out in several ways. After filtration and dissolution of the residue in hydrochloric acid under an atmosphere of carbon dioxide, the ferrous iron may be titrated with potassium dichromate, using sodium diphenylamine sulphonate indicator, or with potassium permanganate [31, 32, 33].

The titration may also be made by a standard ammonium vanadate solution, using *N*-phenylanthranilic acid as indicator [34].

The oxide residue may also be treated by the cold acid decomposition/ammonium vanadate method described in detail earlier, titrating the excess of vanadate with ferrous ammonium sulphate, using sodium diphenylamine sulphonate as indicator [28].

In another method, the residue from the bromine–methanol leach was dissolved in hydrochloric acid under carbon dioxide, and the ferric iron titrated with EDTA at *p*H 2 using salicylic acid as indicator; the solution is then oxidized with ammonium peroxydisulphate and again titrated with EDTA to give the ferrous iron [29].

When metallic iron is measured by the volume of hydrogen evolved by an acid, and iron sulphide by iodometric titration of hydrogen sulphide produced in the same reaction, ferrous oxide has been determined in the remaining solution by titration with potassium permanganate [41].

If silver thiocyanate is used to remove metallic iron, silver in the residue can be dissolved by a mixture of bromine and potassium bromide, and in the silver-free residue ferrous iron can be determined by solution in hydrochloric acid and titration with potassium permanganate or dichromate [42].

After the removal of metallic iron by a solution of ferric chloride, the residue can be decomposed by hydrochloric acid and ferrous iron determined by the conventional permanganate or dichromate titrations [44], by potentiometric titration with potassium bromate [46], or by titration with Complexon III at *p*H 1·5–2 in a nitrogen atmosphere [47].

When metallic iron is removed by 10% potassium chlorate in 52% nitric acid, ferrous oxide in the residue can be determined by dissolving in a mixture of hydrochloric acid, phosphoric acid, and an excess of potassium dichromate. The latter, after addition of diphenylamine indicator, is then titrated with a standard solution of ferrous ammonium sulphate [52].

For samples containing no metallic iron, a number of publications on the determination of ferrous iron offer a wide choice of decomposition techniques, titration, colorimetric or other methods, and detailed instructions for the analysis. The usual solution in hydrochloric and sulphuric acids, preferably in an inert atmosphere, with and without the addition of hydrofluoric acid, followed by titration in sulphuric–phosphoric–boric acid solution with potassium dichromate using diphenylamine indicator, or with potassium permanganate alone, is favoured by many [60, 61, 62, 63, 64, 65].

Several workers have suggested the modification of including an excess of the oxidizing titrant in the decomposition mixture, followed by the titration of this excess [66, 67, 68].

Oxidation of the ferrous iron with a known excess of ammonium vanadate followed by titration of this excess with a standard ferrous salt is recommended by several [69, 70, 71].

For difficultly-decomposable minerals, fusion with a mixture of 21 parts sodium fluoride and 31 parts boric acid, followed by acid solution under carbon dioxide, and final titration with vanadate using phenylanthranilic acid as indicator, has been suggested [71].

Titration of ferrous iron by standard ceric sulphate has been proposed, using as indicator methyl red [72], or *N*-phenylanthranilic acid [73].

A procedure for determining ferrous and ferric iron when they are present together, by an EDTA titration, has been published [74]. Adjust the sample solution containing ferrous and ferric iron, up to a total of 30 mg, to *p*H 1·8–2·2. Dilute to 50–60 ml with previously boiled distilled water. Cover the solution with a 10-mm layer of petroleum ether. Add some sulfosalicylic acid and titrate ferric iron with EDTA in the cold to a red-purple to light yellow endpoint. Agitate the solution vigorously during the titration. Add 0·05–0·1g of ascorbic acid and 10 ml of hexamethylenetetramine solution to attain *p*H 5–6·5. Add some methylthymol blue, and titrate ferrous iron with EDTA to a blue to yellow endpoint.

Colorimetric methods, rather than titrimetric, are favoured by some investigators, especially when the ferrous iron is low. The *o*-phenanthroline [75, 76], 2,2^1-bipyridine [73, 77], and bathophenanthroline [78, 79] procedures for ferrous iron have been described.

One using bathophenanthroline is typical. To the sample or aliquot add 10 ml of 10% ammonium dihydrogen phosphate to complex ferric iron, and adjust the *p*H with 3M sodium acetate to 2·0–2·1. Transfer the solution to a separatory funnel, rinse with bathophenanthroline solution, and add the rinsings to the funnel. Shake, allow to stand, add 10 ml chloroform, shake, allow the layers to separate, and drain the organic layer to a 25-ml volumetric flask. Make to volume with absolute ethanol, and measure the absorbance at 540 nm. The ratio of ethanol to chloroform should be about 1:1; otherwise, the chloroform layer does not separate readily. Too high a concentration of the ammonium dihydrogen phosphate results in the inhibition of the ferrous–bathophenanthroline complex.

A different approach is utilized by digestion of the sample in sulphuric–phosphoric acid, and absorption of the evolved sulphur dioxide in excess potassium dichromate; the latter is titrated with a standard ferrous salt. The principle of this method is that ferrous oxide is dissolved and oxidized with evolution of sulphur dioxide which reduces dichromate [80, 81].

Ferrous iron has been determined by adding an excess of standard silver perchlorate solution and back-titrating with standard potassium bromide, using a silver electrode [82]. An amperometric titration with potassium dichromate [83], and a continuous automatic coulometric analysis [84] have been described for ferrous iron.

For resistant silicates, fusion in an evacuated glass tube, solution of the fusion in an excess of standard iodine monochloride in hydrochloric acid, and titration of the free iodine with standard potassium iodate has been recommended [85].

A modification of this, for silicates which are completely attacked by hydrochloric acid, has been published [86]. Carbon tetrachloride or chloroform is added to the iodine monochloride/hydrochloric acid solution of the sample in a carbon dioxide atmosphere, and the temperature is maintained at 80–85°C for 30 minutes. The liberated iodine is transferred to the absorption system containing the iodide solution. The organic solvent retains most of the iodine. The organic solvent and the digested system are separated, and titrated with standard sodium thiosulphate. For samples requiring the use of an acid mixture where temperature control is most effective, reagents are added at 0°C, the solution is allowed to return to room temperature, and is titrated with 0·01N potassium iodide.

In another method, decomposition is carried out as usual with sulphuric and hydrofluoric acids, but potassium iodate is added initially. The iodate remaining is treated in cooled solution with iodine, and the liberated iodine is titrated with standard sodium thiosulphate [87].

Another line of attack on the problem of ferrous iron analysis has been proposed, namely a measurement of oxygen. The increase of weight of ferrous oxide on calcining a sample for 2 minutes at 1000–1100°C is determined [88]. The ferrous oxide of magnetite is oxidized, whereas the ferrous oxide of $2FeO.SiO_2$ is unaltered. If higher accuracy is desired, a correction for combustion of carbon is made.

A modification has been published, in which the sample is ignited in a limited measurable volume of oxygen or purified air [89]. The volume of oxygen consumed in converting ferrous to ferric oxide is determined. Interference of sulphur is obviated by measuring the sulphate content before and after ignition, calculating the amount of oxygen used for sulphur oxidation, and subtracting this from the volume of oxygen used.

In another procedure, oxide material is dissolved in hydrochloric acid to which a small amount of ferrous iron has been added. If the sample contains active oxygen due to the presence of iron, manganese, cobalt, nickel, or lead in the trivalent or tetravalent state, part of the ferrous solution is oxidized. The ferrous chloride is finally titrated with standard ceric sulphate solution [90].

In the presence of sulphides, digestion at room temperature for an hour or so with a mixture of 5% bromine in anhydrous methanol will dissolve sulphides as well as metallic iron, leaving oxides in the residue [61, 64]. In the latter, ferrous iron may be determined by any appropriate method.

A number of other recorded observations on the determination of ferrous iron deserve mention. One contribution reported that ferrous iron was extracted by aqueous aluminium chloride [91]. In another paper, however, it is stated that this reagent causes a partial dissolution of ferrous sulphide [92]. In the same communication, it is concluded that, in the presence of organic matter, it is unjustifiable to attach even qualitative significance to the results of a ferrous iron determination [92].

In a study of the error caused by the reducing effect of sulphide ion on ferric iron, when determining ferrous oxide in slag and iron ore, it was found that it is proportional to the amount of sulphide sulphur and of ferric iron in the sample [93]. For a sulphide sulphur of 0·1–0·3% and ferric oxide less than 5%, the error in ferrous oxide is less than 2%; if sulphur is 3·2% and ferric oxide 70%, the ferrous oxide error could vary from 7 to 70%.

It has been recommended, when trivalent arsenic is present, to evaporate to a low volume with 6N hydrochloric acid to volatilize arsenic trichloride. If pentavalent arsenic is present, use 20 ml of 18N sulphuric acid to dissolve the sample. It is also suggested that, in the presence of sulphide sulphur, 5 ml of saturated mercuric chloride be added, the sample dissolved in phosphoric acid, and titrated with ammonium vanadate [94].

A sequential heating device for ferrous oxide has been described [95].

Methods for protecting ferrous iron from oxidation during its determination have been studied. None were found to be effective; even the use of nitrogen as an inert atmosphere gave an error of 4% [96].

One investigator, reviewing the errors produced by organic matter, sulphides, vanadium, titanium (III), and manganese (IV), concluded that no direct method for ferrous iron is satisfactory [97].

The determination of ferrous iron in magnetite and ilmenite can be simplified by a prior separation of these minerals by means of Clerici solution of s.g. 4·2 [62].

A study has been made of the effect of metallic iron, derived from the pulverizing of samples, on the ferrous iron determination [98]. Steel and other alloy contamination from the grinding of rock samples can cause appreciable error in ferrous iron determinations when oxidizing decomposition methods, such as potassium iodate, are used. Metallic iron consumes three times as much standard oxidant in the oxidizing decomposition method as it does in the non-oxidizing decomposition methods, such as sulphuric plus hydrofluoric acids.

3. Ferric Iron

Ferric iron in most materials is found by subtracting the ferrous iron, which is determined directly, from the total iron which is determined on another portion of the sample [99, 100, 101, 102]. This, of course, will give an accurate result only when metallic and sulphide iron are absent.

Pyrite is not decomposed in most acid digestions for ferrous iron and can be determined in the residue. Corrections can be applied to subtract the iron of iron sulphides, together with iron of ferrous oxide, from total iron [103]. In sulphide ores, the addition of stannous chloride will suppress the reaction of ferric iron with hydrogen sulphide on acid addition [104]. By running two sets of determinations with and without stannous chloride, the true values of ferrous and ferric iron can be calculated.

A rapid method for ferrous and ferric iron in solutions derived from leaching, refining, or pickling operations has been described [105, 106]. Ferrous iron is titrated with potassium permanganate, or with potassium dichromate in the presence of barium diphenylamine sulphonate indicator. Then a slight excess of copper metal powder is added, the mixture stirred thoroughly, filtered, washed, and the filtrate again titrated with the standard permanganate or dichromate. This titration gives the total iron, and total iron — ferrous iron = ferric iron. Finely divided copper rapidly reduces iron at room temperature, 0·1 g of copper easily reducing 0·1–0·2 g of iron in 2 minutes. In the presence of considerable copper the colour change is slightly different from that occurring in pure solutions, but is unmistakable and reproducible. It is the writer's experience that dichromate is preferable to permanganate for this titration.

For difficultly-decomposed rocks, a polarographic procedure has been published for ferrous and ferric iron [107, 108]. The sample is melted for 6 hours at 1000°C under an inert gas with sodium fluoborate or a flux consisting of 21 parts sodium fluoride to 31 parts boric acid. The melt is dissolved in 1:10 sulphuric acid, sodium oxalate is added as the supporting electrolyte, and ferrous and ferric iron are determined polarographically.

Ferric iron, of course, may be determined by direct titration with a standard solution of titanous sulphate or titanous chloride. Potassium thiocyanate is used as indicator, the endpoint being the disappearance of the red colour of ferric thiocyanate [57]. Titanous salts have been used for this purpose for rocks and minerals [109], mattes and slags [110], and various materials [111].

A titrimetric method for ferric iron has been described. Dissolve 0·1–0·2 g sample in 3 ml hydrofluoric acid and 1 ml sulphuric acid on a hot plate for 3·5 minutes, and transfer to a cooled titrating solution. The latter is made by mixing 10 ml of saturated boric acid, and 50 ml of titanium solution; add 25 ml sulphuric acid, dilute to about 160 ml, boil, cool, add 15 drops of leuco methylene blue and titrate with an automatic titrator, using the coulometric generation of titanous ion as the titrant. The instrument shuts off automatically when the blue colour is reduced to colourless at the endpoint.

The titanium solution is made by dissolving 40 g pure titanium in 1200 ml of 5M sulphuric acid and 100 ml of fluoboric acid, diluting to 2 litres and adding 1 ml of hydrochloric acid. The titanium is oxidized by adding potassium permanganate solution to a persistent pink, and destroying the excess with a little sodium azide.

If the sample contains magnetite, add, in addition to hydrofluoric and sulphuric acids, 1–2 ml of hydrochloric or hydrobromic acid. Determine total iron in the same manner, except that hydrochloric acid must be used for magnetite, and ferrous iron is oxidized by a potassium permanganate solution, later destroying the excess with sodium azide, before titration. Ferrous iron is obtained by subtracting ferric iron from total iron. Pyrite and pyrrhotite interfere by reduction of ferric iron [109].

Solvent extraction has been used to determine ferric iron [112]. Heat the sample in a polythene bottle for 30 minutes with a mixture of sulphuric and hydrofluoric acids, add boric acid, and extract ferric iron with isopropanol. Evaporate an aliquot to drive off hydrofluoric acid, dissolve in sulphuric acid, and determine ferric iron colorimetrically by the *o*-phenanthroline method.

It has been reported that when the temperature is below 80°C the reduction of ferric iron by organic matter is negligible [113].

The determination of ferrous and ferric iron in hydrochloric acid solution stronger than 8M has been carried out by oxidizing iron with potassium ferricyanide, and determining the ferric iron titrimetrically or potentiometrically [114].

A coulometric determination of trivalent iron has been described [115]. Ferric iron is reduced to ferrous if a negative potential is established at the surface of the sample; the current passing through the sample corresponds to the amount of ferric iron. The electrolyte is 1N sulphuric acid, and argon is passed through the cell.

Controlled potential coulometry has been used for the determination of ferric and ferrous iron in slag [116].

In a copper etching bath, pyrophosphate has been shown to be a suitable medium for the simultaneous determination of ferric and ferrous ions by pulse polarography [117].

4. Iron sulphides

The interference of sulphides in the determination of ferrous and ferric iron, and the removal of the former by a solution of bromine–alcohol, have already been discussed.

In a mixture of metallic iron, iron oxides, and ferrous sulphide, the latter has been determined by iodometric titration of hydrogen sulphide evolved upon addition of hydrochloric acid [41, 103, 118].

Treatment of high-sulphur iron ores [61], copper slag [64], and copper mattes and slags [110] with 5% bromine in methanol for 1 hour at room temperature dissolved the sulphides of iron and copper, leaving the oxides in the residue.

It has been stated that for complete extraction of ferrous sulphide from mattes with an ethanol solution of bromine, two treatments are necessary, whereas with a methanol solution a single treatment is sufficient [119].

Several publications discuss the determination of pyrite. For soil, a Soxhlet extraction for 15 hours with 20% hydrochloric acid and then for another 15 hours with 68% nitric acid, leaves pyrite in the residue. Recoveries from known pyrite additions were excellent [120].

In pyrophyllite, the carbonates, oxides, and hydroxides of iron were removed by stirring with 1:1 hydrochloric acid for 90 minutes. Pyrite was then dissolved in a mixture of 30% hydrogen peroxide and 1:1 hydrochloric acid [121].

A method for pyrite in rocks involves decomposition with a mixture of sulphuric and hydrofluoric acids in an atmosphere of carbon dioxide, and separation by filtration of the residue containing the pyrite. Ignition of the residue at 800°C, fusion with potassium pyrosulphate, and solution in dilute sulphuric acid, is followed by determination of the pyritic iron [122].

For shales, a determination to give ferrous sulphide and pyrite has been described [123]. The sample is boiled with hydrochloric acid in a vacuum system, and hydrogen sulphide is swept by a current of carbon dioxide into a solution of potassium hydroxide. A known excess of acidic potassium permanganate is added, followed by an excess of sodium oxalate, and final titration with standard potassium permanganate to give a measure of ferrous sulphide. The residue left after the evolution of hydrogen sulphide with hydrochloric acid contains the pyrite; filtration, solution of the precipitate in acids, and determination of iron gives the content of pyrite.

On mine and concentrator products a differentiation of pyrite from pyrrhotite is sometimes required. A useful procedure, depending on the solubility of pyrrhotite and the insolubility of pyrite in a hot solution consisting of two volumes of water and one of hydrochloric acid, has been described [124]. It must be emphasized that such a differentiation depends on the mineralization of the ore, and the analyst may find it necessary to introduce modifications for the samples of interest.

Heat for 10 minutes 1–2 g of the −100 mesh sample with 50 ml of a solution containing 2 volumes of water to 1 volume of hydrochloric acid. Filter, wash, and determine iron in the filtrate by any convenient method; this represents the iron derived from pyrrhotite. To the precipitate in a beaker add a few crystals of potassium chlorate, 10 ml nitric acid, boil, add 10 ml 1:1 sulphuric acid and evaporate to strong fumes of the latter. Cool, add 5 ml hydrochloric acid, 50 ml water, boil, cool, and determine iron by any appropriate procedure; this represents the iron derived from pyrite.

Filtration is expedited for siliceous samples by using a Gooch crucible, lined with an asbestos pad, in a suction flask. In the original publication, it was stated that magnetite, when present in amounts not exceeding 3% might be considered as being entirely soluble in this solution of 2 water : 1 hydrochloric acid. The writer has found that magnetite in many ores is completely soluble when present in amounts greatly exceeding 3%.

When magnetite is present, of course, the result for pyrrhotite must be corrected. This is done by determining magnetite in a separate sample. To a 1-g sample add a few drops of bromine and 10–15 ml nitric acid, and digest until all sulphides have gone into solution. Add successive quantities of water, and decant until all the iron in the solution has been removed. Add 50 ml 1:1 hydrochloric acid, heat until the magnetite goes into solution, and determine iron by any convenient method. By subtracting this result from that obtained for the iron derived from the pyrrhotite separation the true value for pyrrhotite may be obtained.

Chalcopyrite is nearly insoluble in the 2 water : 1 hydrochloric acid solution and will be present in the residue with the pyrite. If it is the only copper mineral in the sample, the iron it contributes to the pyrite residue may be calculated from the copper result on the basis that chalcopyrite contains 34·5% copper, 30·5% iron, and 35% sulphur. A correction for chalcopyrite is usually unnecessary, except for copper concentrates [124].

5. Magnetite

A knowledge of the magnetite content of mattes and slags is important in copper metallurgy. There are several methods used to isolate and determine this form of iron, which may be present with iron sulphide, iron oxides or silicates, and metallic iron.

One procedure depends on the rapid dissolution of sulphides in a hot nitrochlorate solution, and the relative insolubility of magnetite in the same conditions [125, 126, 127]. It is essentially a method for the determination of ferric iron, the magnetite content being calculated from the ferric iron value on the assumption that all the ferric iron present is in the form of magnetite.

For mattes, weigh 0·5–1 g of –100 mesh sample into a 250 ml beaker, moisten with water, place the beaker in a cooling trough and add 25–50 ml of a nitrochlorate mixture [125]. The latter is made by dissolving 30 g sodium chlorate in hot water, cooling, adding 30 ml nitric acid, cooling and adding 30 ml of 85% sulphuric acid. When the action of the nitrochlorate solution on the sample has subsided, heat the beaker to boiling, add 50 ml cold water, filter, and wash. Determine ferric iron by titrating with a solution of titanous sulphate or chloride, using 5 ml of 10% potassium thiocyanate as indicator; the endpoint is the disappearance of the red colour. $Fe \times 1{\cdot}382 = Fe_3O_4$.

For slags, to 0·5–1 g of –100 mesh sample add 20–40 ml of a hot mixture of equal parts nitric acid and 15% sodium chlorate, boil for 1 minute, and add 100 ml of cold water [126]. Filter on a Gooch crucible and wash with hot water. Transfer to a beaker containing a boiling solution of 7 ml water and 10 ml of a mixture of 3 parts hydrochloric acid and 1 part hydrofluoric acid. Cover and continue boiling until the residue has gone into solution, transfer to a larger beaker containing enough cold water to bring the temperature of the solution to 20°C. Add 20 ml of hydrochloric acid saturated with boric acid, and titrate the ferric iron with standard titanous solution.

A variation of this procedure for copper–nickel blast furnace mattes has been published [127]; mixtures of nitric acid with either potassium chlorate or bromine are allowed to stand in the cold for 16 hours. Results are comparable with the method described above.

Another method for magnetite in copper mattes and slags has been described [128]. The outline below contains additional information kindly supplied by the author in a personal communication. Weigh 1 g into a flask, pass in a stream of carbon dioxide, add an excess of standard 2·5% stannous chloride solution and 50 ml hydrochloric acid. Boil for 20 minutes to remove all hydrogen sulphide, add sufficient hydrofluoric acid to complete the decomposition, add 2–3 g boric acid, dilute with air-free water, and cool. Add marble, 1 g potassium iodide, starch solution, and titrate the excess stannous chloride with 0·1N iodine solution. One ml of the latter is equivalent to 11·577 mg Fe_3O_4. Results are satisfactory in the absence of metallic iron or copper, Fe_2S_3, CuS, Cu_2S, and CuO. The method can be used for nickel slags but not mattes.

A third procedure for magnetite in copper mattes and slags uses a bromine–methanol mixture to dissolve metallic iron and iron sulphide, leaving the oxides entirely in the residue [110]. To 1 g of −150 mesh material add 80 ml of anhydrous methanol and 20 ml bromine. Shake for 1 hour, filter, and wash with methanol. Dissolve the residue in hydrochloric acid, and hydrofluoric acid if necessary, under an atmosphere of carbon dioxide. Determine the ferric iron by titrating with a titanous salt; this represents the magnetite in the sample. A small quantity of copper, in an oxide form, remains in the residue after extraction but has no measurable effect on the iron titration.

Magnetite has been determined in the presence of chalcopyrite by heating and stirring for 5 minutes in 1:1 phosphoric acid; about 1% of the iron in chalcopyrite also dissolves [129].

In the presence of hematite, magnetite has been selectively dissolved by a mixture of phosphoric and tartaric acids, to which an oxidizer such as potassium permanganate or hydrogen peroxide has been added [130].

Magnetite in the presence of iron carbonates and metallic iron has been determined by the selective dissolution of the latter two in ferric chloride solution [131]. The hydrochloric acid formed by hydrolysis of the ferric chloride destroys the adhesion of the magnetite with iron carbonates while dissolving both the latter and metallic iron.

In silicate minerals containing ferrous iron, magnetite has been determined by treatment for two hours at 50°C with 2·5N nitric acid, filtration, and washing. The residue, containing the magnetite, is dissolved in sulphuric and phosphoric acid in an atmosphere of carbon dioxide, and the ferrous iron titrated in the usual manner [132].

In products of reduction roasting, containing metallic iron, wustite, fayalite, and ferrous oxide, magnetite has been selectively dissolved in a boiling alcoholic solution of stannous chloride [133].

6. Miscellaneous Iron Compounds

Hematite and iron hydroxides have been determined by selective dissolution [134]. Magnetite is first separated magnetically, and the remainder of the sample is treated with an acid mixture containing 0·5N citric acid, 0·4N hydrochloric acid, and 0·01N ferrous ammonium sulphate, for 4 hours at 100°C with periodic stirring. The sample is filtered, ignited, and hematite iron is determined in this residue. Iron hydroxides are found from the difference between hematite and total non-magnetic iron.

Several publications discuss the chemical analysis of carbide, oxide and nitride phases in steels, alloys, and iron powders by anodic dissolution and other electrolytic techniques [135, 136, 137, 138].

The phase analysis of iron in ores and compounds has been reviewed in several papers [139, 140, 141].

REFERENCES

1. Petzold, F., *Arch. Eisenhuttenw.* **12**, 237–43 (1938). *C.A.* **33**, 497 (1939).
2. Lukina, Z. E., *Zavod. Lab.* **9**, 19–22 (1940). *C.A.* **35**, 7872 (1941).
3. Morris, J. P., *U.S. Bur. Mines, Rept. Investigations* 3824, 1945.
4. Murata, Y., and Kasaoka, S., *Bunseki Kagaku* **5**, 156–9, (1956). *C.A.* **51**, 9411 (1957).
5. Kondo, A., and Fuke, Y., *Nippon Kinzoku Gakkaishi* **22**, 286–90 (1958). *C.A.* **55**, 15221 (1961).
6. Murata, Y., and Kasaoka, S., *Bunseki Kagaku* **7**, 721–4 (1958). *C.A.* **53**, 19692 (1959).
7. Murata, Y., and Kasaoka, S., *Bunseki Kagaku* **7**, 50–2 (1958). *C.A.* **54**, 168 (1960).
8. Tkachenko, N. G., Davidenko, P. I., and Dobrzhanskii, A. V., *Zavod. Lab.* **29**, 536–8 (1963). *C.A.* **59**, 4535 (1963).
9. Stognii, N. I., *Zavod. Lab.* **28**, 1068 (1962). *C.A.* **58**, 11948 (1963).
10. Aubry, J., and Perrot, P., *Chim. Anal.* **47**, 177–9 (1965). *C.A.* **63**, 3613 (1965).
11. Easton, A. J., and Lovering, J. F., *Geochim. Cosmochim. Acta* **27**, 753–67 (1963).
12. Williams, C. E., Barrett, E. P., and Larsen, B. M., *U.S. Bur. Mines Bull.* 270, 1927.
13. Budnikov, P. P., and Mel'nikova, A. N., *J. Applied Chem.* (U.S.S.R.) **13**, 1732–8 (1940). *C.A.* **35**, 3922 (1941).
14. Sosnovskii, B. A., *Zavod. Lab.* **16**, 872–3 (1950). *C.A.* **45**, 974 (1951).
15. Lavrukhina, A. K., *Trudy Komissii Anal. Khim. Akad. Nauk S.S.R.* **3**, 267–80 (1951). *C.A.* **47**, 2632 (1953).
16. Vojtiskova, V., *Hutn. listy* **23**, 276–9 (1968). *C.A.* **69**, 24299 (1968).
17. Uno, T., Kawamura, K., and Kitayama, M., *Fuji Sietetsu Giho* **11**, 25–34 (1962). *C.A.* **58**, 3874 (1963).
18. Riott, J. P., *Ind. Eng. Chem. Anal. Ed.* **13**, 546–9 (1941).
19. Wesly, W., *Chem.-Ztg.* **68**, 106–7 (1944). *C.A.* **39**, 37 (1945).
20. Babenko, N. L., and Terzeman, L. N., *Tr. Inst. Met. Obogashch., Akad. Nauk Kaz. S.S.R.* **31**, 66–71 (1968). *C.A.* **72**, 28096 (1970).
21. Vorlicek, J., and Lang, S., *Tech. Zpravodaj Zelezorudnych Dola Hrudkoven* **2**, (6), 19–28 (1963). *C.A.* **62**, 3393 (1965).
22. Prasad, T. P., *Talanta* 16, 1484–5 (1969).
23. Fedorov, A. A., *Sb. Tr. Tsent. Nauch.-Issled. Inst. Chern. Met.* No. 73, 75–9 (1969). *C.A.* **73**, 62351 (1970).
24. Habashy, M. G., *Anal. Chem.* **33**, 586–8 (1961).

25. Kudelin, V. N., *Byull. Nauch.-Tekh. Inform. Gosudarst. Nauch.-Issledovatel. i Proekt. Inst. po Obogashchesii i Aglomeratsii Rud Chern. Metallov* **1959**, 117–24. *C.A.* **55**, 25587 (1961).
26. Vogel, H. U., *Arch. Eisenhuttenw.* **20**, 287–92 (1949). *C.A.* **44**, 71 (1950).
27. Okura, Y., *Nippon Kinzoku Gakkaishi* **24**, 237–41, 241–5, 289–93, 293–5, 296–300 (1960). *C.A.* **61**, 36 (1964).
28. Kraft, G., and Fischer, J., *Z. Anal. Chem.* **197**, 217–21 (1963). *C.A.* **60**, 14 (1964).
29. Wakamatsu, S., *Bunseki Kagaku* **14**, 297–301 (1965). *C.A.* **63**, 9050 (1965).
30. Kinson, K., Dickeson, J. E., and Belcher, C. B., *Anal. Chim. Acta* **41**, 107–12 (1968).
31. Kawamura, K., Watanabe, S., and Sasaki, H., *Nippon Kinzoku Gakkaishi* **32**, 676–81 (1968). *C.A.* **69**, 113240 (1968).
32. Rivlin, I. Y., Totsman, E. A., and Gromozova, I. V., *Tr. Vses. Nauch.-Issled. Inst. Abrazivov Shlifovaniya* **1967**, No. 4, 44–56. *C.A.* **69**, 24293 (1968).
33. Krath, E., *Arch. Eisenhuttenw.* **39**, 49–55 (1968). *C.A.* **68**, 84076 (1968).
34. Leblond, A. M., Wendling, R., and Bourdieu, J. M., *Chim. Anal.* **50**, 431–8 (1968). *C.A.* **71**, 18537 (1969).
35. Sant, B. R., and Prasad, T. P., *Talanta* **15**, 1483–6 (1968).
36. Sasuga, H., Kato, K., and Asaoka, Y., *Bunseki Kagaku* **19**, 542–6 (1970). *C.A.* **73**, 51997 (1970).
37. Moldovan, M., and Horak, I., *Rev. Chim.* **21**, 639–41 (1970). *C.A.* **74**, 82765 (1971).
38. Blum, S. C., and Searl, T. D., *Anal. Chem.* **43**, 150–2 (1971).
39. Malecki, J., and Leszczynski, St., *Przemysl Chem.* **21**, 298–301 (1937). *C.A.* **32**, 2454 (1938).
40. Portessis, M., *Compt. rend.* **228**, 1233–4 (1949). *C.A.* **43**, 8303 (1949).
41. Piksarv, A., Arro, H., and Vares, V., *Tr. Tallinsk. Politekhn. Inst., Ser. A* No. **215**, 137–48 (1964). *C.A.* **63**, 17134 (1965).
42. Marion, F., and Aubry, J., *Chim. Anal.* **41**, 401–7 (1959). *C.A.* **54**, 3072 (1960).
43. Aubry, J., and Marion, F., *Compt. rend.* **235**, 1509–10 (1952). *C.A.* **47**, 5298 (1953).
44. Saunders, H. L., *Iron & Steel Institute (London) Special Rept.* No. **18**, 96–7 (1937). *C.A.* **31**, 8425 (1937).
45. Kitahara, S., *J. Sci. Research Inst.* **43**, 1–27, 133–59 (1949). *C. A.* **43**, 6106 (1949).
46. Neumann, B., and Meyer, G., *Z. Anal. Chem.* **129**, 229–32 (1949). *C.A.* **44**, 71 (1950).
47. Vorlicek, J., and Vydra, F., *Sb. Praci Vzykum. Ustavu ZDHE* **5**, 147–54 (1964). *C.A.* **62**, 12432 (1965).
48. Kashparova, O. D., and Fedorov, A. A., *Sb. Tr. Tsent. Nauch.-Issled Inst. Chern. Met.* No. **73**, 70–5 (1969). *C.A.* **73**, 62354 (1970).

49. Lebedev, O. P., Frantsuzova, T. A., and Kudelin, V. N., *Obogashch. Polez. Iskop* No. **3**, 3–6 (1968). *C.A.* **70**, 111419 (1969).
50. Maynes, A. D., *Chem. Geol.* **6**, 255–63 (1970).
51. Balabanoff, L., Schmidt, K. E., and Seeger, S. B., *Z. Anal. Chem.* **204**, 107–10 (1964). *C.A.* **61**, 10032 (1964).
52. Lim, H. K., Han, S. D., and Chu, Y. A., *Punsok Hwahak* **8**, (4) 15–23 (1970). *C.A.* **75**, 94340 (1971).
53. Lim, H. K., Han, S. D., and Chu, Y. A., *Choson Minjujuui Immin Konghwaguk Kwahagwon Tongbo* **1**, 27–9 (1972). *C.A.* **77**, 83200 (1972).
54. Sumi, K., Hata, T., and Hagiwara, T., *Bunseki Kagaku* **21**, 94–6 (1972). *C.A.* **77**, 13521 (1972).
55. Hillebrand, W. F., Lundell, G. E. F., Bright, H. A., and Hoffman, J. I., *Applied Inorganic Analysis*, 2nd ed., New York, Wiley, 1953.
56. Maxwell, J. A., *Rock and Mineral Analysis*, New York, Wiley-Interscience, 1968.
57. Young, R. S., *Chemical Analysis in Extractive Metallurgy*, London, Charles Griffin, 1971.
58. Wilson, A. D., *Bull. Geol. Survey Gt. Brit.* No. **9**, 56–8 (1955).
59. Wilson, A. D., *Analyst* **85**, 823–7 (1960).
60. Kovtun, M. S., *Zavod. Lab.* **5**, 1042–6 (1936). *C.A.* **31**, 972 (1937).
61. Okura, Y., *Bunseki Kagaku* **9**, 841–52 (1960). *C.A.* **55**, 19596 (1961).
62. Fahey, J. J., *U.S. Geol. Surv., Profess. Papers* No. **424–C**, 386–7 (1961).
63. Schafer, H. N. S., *Analyst* **91**, 763–70 (1966).
64. Tsuruya, S., and Kanno, T., *Tohoku Daigaku Senko Seiren Kenbyusho Iho* **22**, 1–6 (1966). *C.A.* **69**, 92722 (1968).
65. Babenko, N. L., and Mityaeva, P. A., *Tr. Inst. Met. Obogashch., Akad. Nauk Kaz. S.S.R.* **33**, 87–91 (1968). *C.A.* **69**, 56706 (1968).
66. Reichen, L. E., and Fahey, J. J., *U.S. Geol. Surv., Bull.* No. **1144B**, 1962.
67. Spivakovski, V. B., and Zimina, V. A., *Zavod. Lab.* **28**, 290 (1962). *C.A.* **58**, 2848 (1963).
68. Pannani, K., and Agnihotri, S. K., *Lab. Pract.* **15**, 867 (1966).
69. Ishibashi, M., and Kusaka, Y., *J. Chem. Soc. Japan, Pure Chem. Sect.* **71**, 160–3 (1950). *C.A.* **45**, 4600 (1951).
70. Peters, A., *Neues Jahrb. Mineral. Monatsh.* **1968**, 119–25. *C.A.* **68**, 92703 (1968).
71. Novikova, Y. N., *Zh. Anal. Khim.* **23**, 1057–9 (1968). *C.A.* **69**, 83167 (1968).
72. Percs, E., *Magyar Gyogyszeresztud. Tarsasag Ertesitoje* **14**, 456–8 (1938). *C.A.* **32**, 8985 (1938).
73. French, W. J., and Adams, S. J., *Analyst* **97**, 828–31 (1972).
74. Koros, E., and Barcza, L., *Chemist-Analyst* **48**, 69, 72–3 (1959).
75. Shapiro, L., *U.S. Geol. Survey, Profess. Paper* No. **400–B**, 226 (1960).

76. Tachibana, K., *Mem. Fac. Sci. Kyushu Univ. Ser. C.* **4**, 239–46 (1961). *C.A.* **57**, 11850 (1962).
77. Riley, J. P., and Williams, H. P., *Mikrochim. Acta* **1959**, 516–24. *C.A.* **58**, 10709 (1963).
78. Ghosh, M. M., O'Connor, J. T., and Engelbrecht, R. S., *J. Am. Water Works Ass.* **59**, 897–905 (1967).
79. Pollock, E. N., and Miguel, A. N., *Anal. Chem.* **39**, 272 (1967).
80. Seil, G. E., *Ind. Eng. Chem. Anal. Ed.* **15**, 189–92 (1943).
81. Bouvier, J. L., Sen Gupta, J. G., and Abbey, S., *Geol. Surv. Can.*, Pap. No. 72–31 (1972).
82. Ungethuem, H., *Acta Geol. Geogr. Univ. Comenianae, Geol.* **15**, 83–5 (1968). *C.A.* **72**, 38545 (1970).
83. Budilovskii, M. Y., and Usatenko, Y. I., *Zavod. Lab.* **36**, 788–92 (1970). *C.A.* **73**, 126631 (1971).
84. De Kainlis, G., Merigot, D., and Pourcel, C., *Chim. Anal.* **53**, 696–700 (1971). *C.A.* **76**, 94085 (1972).
85. Hey, M. H., *Mineralog. Mag.* **26**, 116–18 (1941).
86. Das Gupta, H. N., and Mitra, N. K., *J. Inst. Chem.* **39**, 250–4 (1967).
87. Van Loon, J. C., *Talanta* **12**, 599–603 (1965).
88. Plotkin, N. Z., Usatenko, Y. I., and Bulakhova, P. A., *Zavod. Lab.* **15**, 999–1000 (1949). *C.A.* **44**, 975 (1950).
89. Habashy, M. G., *Anal. Chem.* **34**, 1015–18 (1962).
90. Van Oosterhout, G. W., and Visser, J., *Anal. Chim. Acta* **33**, 330–2 (1965).
91. Vizier, J. F., and Blanch, C., *Cah. ORSTOM, Ser. Pedol.* **7**, 435–45 (1969). *C.A.* **72**, 139319 (1970).
92. Pruden, G., and Bloomfield, C., *Analyst* **94**, 688–9 (1969).
93. Ikegami, T., and Morita, S., *Nippon Kinzoku Gakkai-Shi* **B14**, No. 6, 52–7 (1950). *C.A.* **47**, 7942 (1953).
94. Ikegami, T., Kammori, O., and Morita, S., *Tetsu-to-Hagane* **39**, 1350–62 (1953). *C.A.* **49**, 7445 (1955).
95. Shapiro, L., and Rosenbaum, F., *U.S. Geol. Surv., Profess. Papers* No. **450–C**, 102–3 (1962).
96. Tselinskii, Y. K., Gorbenko, F. P., and Nikol'skaya, N. N., *Tr. Vses. Nauch.-Issled. Inst. Khim. Reaktivov Osobo Chist. Khim. Veshchestv* No. **26**, 281–4 (1964). *C.A.* **66**, 72149 (1967).
97. Wohlmann, E., *Z. Angew. Geol.* **10**, 663–6 (1964). *C.A.* **62**, 9780 (1965).
98. Ritchie, J. A., *Geochim. Cosmochim. Acta* **32**, 1363–6 (1968).
99. Stognii, N. I., *Zavod. Lab.* **8**, 391–5 (1939). *C.A.* **34**, 45 (1940).
100. Orlova, N. A., *V. Bor'beza Tekh. Progress* **1957**, No. 2, 108. *C.A.* **54**, 13972 (1960).
101. Zalmanzon, E. S., *Litologiya i Polezn. Iskop.* **3**, 138 (1966). *C.A.* **66**, 16312 (1967).
102. Vieux, A. S., and Kabwe, C., *Chim. Anal.* **52**, 866–9 (1970). *C.A.* **74**, 9300 (1971).
103. Rozenovich, V. A., *Zavod. Lab.* **10**, 484–6 (1941). *C.A.* **36**, 56 (1942).

104. USATAYA, E. S., *Vsesoyuz. Nauch.-Issledovatel. Inst. Metrol., Sbornik Trudov* **1941**, No. 3, 30–4. *C.A.* **39**, 5198 (1945).
105. PERCIVAL, J. O., *Ind. Eng. Chem. Anal. Ed.* **13**, 71–2 (1941).
106. YOUNG, R. S., *Chemist-Analyst* **32**, 79 (1943).
107. MIKHAILOVA, Z. M., VARUSHKINA, A. A., MIRSKII, R. V., and SHIL'DKROIT, E. A., *Tr. Kuibyshevsk. Gos. Nauchn.-Issled. Inst. Neft. Prom.* **1963**, (20), 124–9. *C.A.* **61**, 6384 (1964).
108. MIKHAILOVA, Z. M., MIRSKII, R. V., and VARUSHKINA, A. A., *Zh. Analit. Khim.* **18**, 856–8 (1963). *C.A.* **59**, 10759 (1963).
109. CLEMENCY, C. V., and HAGNER, A. F., *Anal. Chem.* **33**, 888–92 (1961).
110. KORAKAS, N., *Trans. Institution Mining Met.* **72**, Part 1, 35–44 (1962–63).
111. AUBRY, J., BOURRET, P., and EVRARD, O., *Chim. Anal.* **45**, 449–51 (1963). *C.A.* **60**, 13 (1964).
112. ABE, Y., and NARUSE, A., *Bunseki Kagaku* **13**, 417–22 (1964). *C.A.* **61**, 2470 (1964).
113. MITSUCHI, M., and OYAMA, M., *Nippon Dojo-Hiryogaku Zasshi* **34**, 23–7 (1963). *C.A.* **60**, 9905 (1964).
114. WISNIEWSKI, W., and BASINSKA, H., *Chem. Anal.* (*Warsaw*) **12**, 99–105 (1967). *C.A.* **67**, 50087 (1967).
115. PIL'KO, E. I., and KOZHEUROV, V. A., *Sb. Nauch Tr. Chelyabinsk Politekh. Inst.* **1970**, No. 66, 67–71. *C.A.* **76**, 54081 (1972).
116. ALTMAN, R. L., *Anal. Chim. Acta* **63**, 129–38 (1973).
117. PARRY, E. P., and ANDERSON, D. P., *Anal. Chem.* **45**, 458–63 (1973).
118. CHERNYI, A. T., and PODOINIKOVA, K. V., *Zavod. Lab.* **16**, 1308–9 (1950). *C.A.* **45**, 10126 (1951).
119. FILIPPOVA, N. A., MARTYNOVA, L. A., and GUSEL'NIKOVA, N. Y., *Zavod. Lab.* **37**, 783–4 (1971). *C.A.* **75**, 104695 (1971).
120. PETERSEN, L., *Acta Agr. Scand.* **19**, 40–4 (1969). *C.A.* **71**, 87388 (1969).
121. SADUOKASOV, A. S., DZHAKUPOVA, G. Z., and ZEL, F. I., *Tr. Khim.-Met. Inst. Akad. Nauk Kaz. S.S.R.* **10**, 152–9 (1969). C. A. **73**, 105211 (1970).
122. TRUSOV, Y. P., ZHUR. *Anal. Khim.* **14**, 139–40 (1959). *C.A.* **53**, 11104 (1959).
123. NEGLIA, S., and FAVRETTO, L., *Clay Minerals Bull.* **5**, 37–40 (1962).
124. MCLACHLAN, C. G., *Am. Inst. Min. Met. Eng. Trans.* **112**, 593–6 (1934).
125. HAWLEY, F. G., *Eng. Mining World* **2**, 270–2 (1931).
126. ROBERTS, L. E., and NUGENT, R. L., *U.S. Bur. Mines Rept. Invest.* 3120 (1931).
127. DRUMMOND, P. R., *Can. Inst. Mining Met. Trans.* XLIII, 627–52 (1940).
128. KINNUNEN, J., *Chemist-Analyst* **40**, 89–92 (1951).
129. MORACHEVSKII, Y. V., and PINCHUK, N. K., *Vestnik Leningrad Univ.* **11**, No. 22, Ser. Fiz. i Khim. No. 4, 170–5 (1956). *C.A.* **51**, 8583 (1957).

130. Pinchuk, N. K., and Morachevskii, Y. V., *Vestnik Leningrad Univ.* **13**, No. 4, Ser. Fiz. i Khim. No. 1, 126–33 (1958). *C.A.* **52**, 12660 (1958).
131. Lebedev, O. P., Frantsuzova, T. A., and Kudelin, V. N., *Zavod. Lab.* **31**, 1069–70 (1965). *C.A.* **63**, 17148 (1965).
132. Fedorova, M. N., and Krivodubskaya, K. S., *Obogashch. Rud* **13**, 61–2 (1968). *C.A.* **71**, 18604 (1969).
133. Frantsuzova, T. A., and Lebedev, O. P., *Sb. Nauch. Tr., Nauch.-Issled. Proekt. Inst. Obogashch. Aglom. Rud. Chern. Metal.* **1969**, No. 10, 233–7. *C.A.* **73**, 94340 (1970).
134. Khait, V. I., *Obogashch. Rud.* **10**, (6), 35–6 (1965). *C.A.* **65**, 4665 (1966).
135. Klyachko, Y. A., Shapiro, M. M., and Yakovleva, E. F., *Tr. Seminara po Zharostoikim Materialam, Akad. Nauk. Ukr. S.S.R., Inst. Metallokeram. i Spets. Splavov, Kiev,* **1960**, No. 6, 59–63. *C.A.* **57**, 47 (1962).
136. Lashko, N. F., and Yakovleva, E. F., *Sb. Tr. Tsent. Nauch.-Issled. Inst. Chern. Met.* **1969**, No. 73, 139–44. *C.A.* **73**, 83455 (1970).
137. Chladek, O., *Hutn. listy* **26**, 590–2 (1971). *C.A.* **75**, 115438 (1971).
138. Bulanov, V. Y., Mokshantsev, G. F., Mullin, A. K., and Sinyukhin, A. V., *Metody Anal. Issled. Svoistv. Mater.* **1971**, 115–20. *C.A.* **77**, 121789 (1972).
139. Uno, T., Kawamura, K., and Kitayama, M., *Fuji Seitetsu Giho* **11**, 476–87 (1962). *C.A.* **60**, 1105 (1964).
140. Fedorova, M. N., Kostousova, T. I., and Krivodubskaya, K. S., *Tr. Nauchn.-Issled. Proekt. Inst. Obogashch. Mekhan. Obrabotski Polez. Iskop. Uralmekhanobor* No. **12**, 294–309 (1965). *C.A.* **66**, 72108 (1967).
141. Filippova, N. A., and Martynova, L. A., *Sb. Nauch. Tr., Gos. Nauch.-Issled. Inst. Tsvet. Metal.* No. **28**, 72–85 (1968). *C.A.* **70**, 43637 (1969).

Lead

Lead is found in nature as sulphide and various oxidized forms; in the different processes of extractive metallurgy the element may occur in the sulphide, oxidized, or metallic state. Phase analysis of lead is important, and a number of publications on this subject have appeared.

1. Slags

A useful procedure for determining the forms of lead in slag and similar materials has been employed for many years by the author [1]. To 0·5–5 g of –100 mesh sample, add 25 ml of a saturated solution of ammonium acetate, dilute to 100 ml, and boil for 10 minutes. Filter through an asbestos mat on a Gooch crucible and wash with hot water. The filtrate contains the lead which was present as oxide, sulphate or basic sulphate; the residue contains lead sulphide, silicate, and metal.

Transfer the residue to a beaker and evaporate nearly to dryness. Add 10–20 ml of 10% silver nitrate solution and allow to stand for an hour with occasional stirring. Filter through asbestos on a Gooch crucible and wash thoroughly. The filtrate contains the lead that was originally present in the metallic form; the residue contains the lead present as sulphide and silicate.

Transfer this residue to a beaker, evaporate nearly to dryness, and add 25–50 ml of a saturated solution of sodium chloride containing 60 g of $FeCl_3.6H_2O$ per litre. Allow to stand for 12 hours, with occasional agitation. Filter the solution through filter paper, wash the residue once with some of the original solvent, and then wash thoroughly with hot water. The filtrate contains the lead that was originally present as sulphide; the final residue is lead silicate. The filtrates and residue are analysed for lead by any convenient procedure.

In another method for lead compounds in slags, "oxidized" lead is dissolved by boiling the sample in 15% ammonium acetate solution for 30 minutes [2]. From the residue, metallic lead is taken into solution by boiling with 5% copper nitrate for 10 minutes, then lead sulphide is dissolved by boiling with 4N sodium chloride/3N hydrochloric acid for 1·5 hours. "Difficultly-soluble" lead compounds remain in the final residue.

2. Ores and Concentrator Products

A method to differentiate various forms of lead in complex ores has proved useful to the author on numerous occasions [3]. To 1 g of −200 mesh sample in a stoppered Erlenmeyer flask add 100 ml of 60% ammonium acetate solution. Allow to stand 1·5 hours, shaking occasionally. Filter through Whatman No. 42 paper and wash thoroughly by decantation, retaining as much of the residue in the flask as possible. The filtrate contains lead which was originally present as sulphate, carbonate, or oxide; the precipitate represents lead present as sulphide, phosphate, and vanadate.

Replace the original flask under the funnel and pour through the paper in successive small increments a total of 100 ml of cold 10% by volume perchloric acid. Lead sulphide is virtually insoluble in this reagent, whereas the phosphates and vanadates will dissolve. Reserve the lead sulphide residue on the paper for addition to the residue in the original flask. Stopper the flask and allow to stand for 2 hours, shaking occasionally. Filter through a Whatman No. 42 paper and wash thoroughly with cold 1% perchloric acid. The lead phosphate and vanadate are in the filtrate; lead sulphide remains on the paper and can be combined with the small quantity which was left on the paper from the first filtration.

If the lead minerals are finely disseminated in a siliceous gangue, add a few drops of hydrofluoric acid to the sample during both leaches. Small quantities of hydrofluoric acid are without effect on lead sulphide; as much as 5% hydrofluoric acid by volume dissolves insignificant amounts with this procedure.

Finely divided lead sulphide has a tendency to oxidize on exposure to a combination of air and moisture; filtrations should therefore not be unduly delayed.

There is no explosion hazard when the cold, dilute perchloric acid solution is filtered through paper. The residue of sulphide on the paper should be evaporated several times with nitric acid to destroy the filter paper and all organic matter, before proceeding with the determination of lead by any appropriate method. There must, of course, be no attempt to shorten the treatment period by using a warm, stronger perchloric acid solution.

A number of other methods for phase analysis of lead minerals and ores have been proposed. In a mixture of $PbSO_4$, $PbCO_3$, $Pb_5Cl(PO_4)_3$, PbS, $PbCrO_4$, and $Pb_5(VO_4)_3Cl$, treatment with 25% sodium chloride dissolves all the lead sulphate and 0·4–0·7% of the lead carbonate.

When the residue is reacted with 200 times as much 15% ammonium acetate, the lead carbonate dissolves, together with 0·3% of $Pb_5Cl(PO_4)_3$ and only 0·05% of lead sulphide. Treatment of this second residue with 2% sodium hydroxide solution dissolves the lead chromate and 0·3–0·4% of the $Pb_5Cl(VO_4)_3$. When the third residue is reacted with a mixture of 25% sodium chloride and 0·5% hydrochloric acid, the solution contains $Pb_5Cl(PO_4)_3$, $Pb_5Cl(VO_4)_3$ and possibly up to 6% lead sulphide [4].

Other workers have dissolved anglesite, lead sulphate, in 25% sodium chloride solution, followed by cerussite, lead carbonate, in a solution of 15% ammonium acetate and 3% acetic acid [5, 6, 7, 8]. Galena can be dissolved from the second residue by a solution containing 25% sodium chloride and 6% ferric chloride, or by dilute acetic acid to which hydrogen peroxide has been added.

The following detailed description is typical [5]. Add 25% sodium chloride solution to the sample, stir, and allow to stand for an hour. Filter, wash, and determine in the filtrate the lead from anglesite. Treat the precipitate with a solution of 15% ammonium acetate containing 2% acetic acid, filter, and wash. The lead from cerussite is in the filtrate. Treat the residue with a solution of sodium hydroxide, filter, and retain the filtrate. Wash the precipitate with water, and mix with ammonium acetate solution. Filter the mixture and combine this filtrate with the previous one. Determine lead of crocoite and wulfenite in this filtrate. Treat the precipitate with sodium chloride solution acidified with hydrochloric acid. Filter and determine lead of vanadinite and pyromorphite in the filtrate. Wash the precipitate with sodium chloride solution, treat with a solution of ferric chloride, filter, and wash the precipitate with water. Determine lead of plumbojarosite, beaverite and beudantite in the precipitate. Determine lead of galena in the filtrate.

A similar representative method gives the following directions [6]. Stir the sample for 1 hour with 50–100 ml of 25% sodium chloride solution. Filter, wash, and determine lead from anglesite in the filtrate. Stir the precipitate for 45 minutes with 50–100 ml of 15% ammonium acetate containing 20 ml/litre of acetic acid. Filter, wash with warm 5% ammonium acetate solution, and determine in this filtrate the lead from cerussite. In the presence of mimetite and absence of wulfenite, stir the precipitate for 20 minutes with 50–100 ml of 25% sodium chloride solution containing 5 ml/litre of hydrochloric acid and 5 mg of mercury as mercuric nitrate. Filter and wash with 2% sodium chloride solution. Determine lead of mimetite, vanadinite, crocoite, and pyromorphite in the filtrate. Stir the precipitate for 40 minutes with 50 ml of 30% hydrogen peroxide at 60–70°C, and afterwards for 15 minutes at the same temperature with 50 ml of 40% ammonium acetate containing 3% acetic acid. Filter and wash with water. Determine lead of galena in the filtrate, and lead of plumbojarosite, beaverite, and beudantite in the precipitate.

In the absence of mimetite, stir the precipitate from the separation of cerussite for 3 hours with 50–100 ml of 2% sodium hydroxide solution. Filter, wash, and retain the filtrate. Stir the precipitate for 30 minutes with 50 ml of ammonium acetate solution, filter, and wash with 5% ammonium acetate. Determine lead of crocoite and wulfenite in the combined filtrates. Continue with the precipitate as given above in the presence of mimetite [6].

In another analytical scheme for ores, the following succession of leach solutions is employed: sodium chloride to remove anglesite; ammonium acetate to dissolve cerussite; sodium hydroxide to extract crocoite, lead chromate; a mixture of sodium chloride and hydrochloric acid to solubilize pyromorphite, lead chlorophosphate, and vanadinite, lead chlorovanadate; and finally a mixture of sodium chloride and ferric chloride to dissolve galena [9].

Lead dusts have been subjected to phase analyses. Lead oxide is removed in acetic acid, lead sulphate in sodium chloride solution, metallic lead in copper nitrate solution, lead arsenate by acidified sodium chloride solution, and lead sulphide by a mixed solution of sodium chloride and ferric chloride [10].

It has been found, when anglesite and cerussite are determined in the presence of galena and pyrite, that pyrite increases the solubility of galena. The effect can be suppressed by adding ascorbic acid and cupric chloride. To 0·5 g of ore add 100 ml of a solution containing 15 wt. % of ammonium acetate, 3 wt. % of acetic acid, 1 wt. % of ascorbic acid, and 0·25 mg cupric chloride. Stir for 1 hour, add 1 wt. % of gelatin, and determine lead polarographically [11].

In a study of Trilon in lead phase analysis, it has been found that Trilon dissolves anglesite, cerussite, crocoite, pyromorphite, and wulfenite, but dissolves galena only after the anionic part of the crystal lattice is destroyed by treatment with hydrogen peroxide and Trilon [12].

When finely ground galena intimately associated with, or bound to, various minerals was treated with a solution of sodium chloride and ferric chloride, the free and partially loosened galena dissolved. An insoluble residue of galena was bound firmly to pyrite, sphalerite, and certain gangue minerals, and even a mixture of ammonium acetate and hydrogen peroxide did not liberate this galena [13].

3. Metallic Lead

Several investigations have discussed the determination of metallic lead in the presence of other forms of this element. One recommends adding to the sample an excess of 0·5N silver nitrate solution, stirring, and allowing to stand for 15 minutes. Filter, wash with 5% acetic acid, acidify the filtrate with nitric acid, and titrate the excess silver nitrate with 0·25N ammonium thiocyanate solution, using ferric alum as indicator [14].

It has been reported that metallic lead in lead oxide may be measured by treating the sample in a nitric acid solution containing an excess of potassium permanganate, and titrating the latter with a standard solution of ferrous ammonium sulphate [15].

Metallic lead in the presence of lead sulphide has been found by stirring the sample for 1 hour in a 5% copper nitrate solution, filtering, and washing; metallic lead is in the filtrate and lead sulphide in the residue [16].

4. Lead Oxide

A few publications have appeared on the analysis of lead oxide in the presence of other phases of this element. In one, an effective separation of lead monoxide from metallic lead, lead dioxide, and lead sulphate was obtained by treatment for 20–30 minutes with 3 acetic acid : 1 water at room temperature, under an atmosphere of carbon dioxide [17].

A different approach has been recommended for lead oxide in the presence of lead. Heat 1 g in 5 ml mercury, cool, add 20 ml of 27% ammonium chloride, and steam-distil off the ammonia liberated in the reaction: $PbO + 2NH_4Cl \rightarrow PbCl_2 + H_2O + 2NH_3$. Collect the ammonia in boric acid and titrate with 0·1N hydrochloric acid, using bromocresol green + methyl red as indicator [18].

A polarographic determination of oxidized lead in the presence of metallic lead has been outlined. Dissolve 1 g in 10 ml of mercury in the presence of a flowing butane–propane mixture. Oxide compounds are then dissolved in 50 ml of 0·4N hydrochloric acid. At 15–20-minute intervals take from this solution 5-ml portions and analyse polarographically, either in the original solution or after addition of 0·5 ml of 5% sodium hydroxide. Plot on a diagram the quantities of lead versus time. Find the quantity of oxidized lead originally present by extrapolating the straight line obtained to time zero at which hydrochloric acid was added to the amalgam [19].

A rapid method for the determination of the degree of oxidation of lead powders has been proposed. Digest 20 g with 250 ml of hot 5% acetic acid, and mix until the residue collects into a spongy globule of lead. Decant the solution, compress the globule in a press at 150 kgf/cm² (14710kN/m², 147 bars), dry the disc with filter paper, and weigh. The time for this estimation is stated to be 5 minutes, and the error not greater than ±0·6% [20]. The quoted time must surely exclude the digestion period, and the figure for percentage error seems remarkably low for this type of rapid test.

The phase analysis of chlorination roasting products for lead and other metals has been discussed [21].

REFERENCES

1. OLDRIGHT, G. L., and MILLER, V., *U.S. Bur. Mines Rept. of Invest.* 2954 (1929).
2. TROFIMOVA, S. G., *Izv. Akad. Nauk Uz. S.S.R., Ser. Tekhn. Nauk* **6**, 75–84 (1962). *C.A.* **57**, 4022 (1962).
3. YOUNG, R. S., GOLLEDGE, A., and TALBOT, H. L., *Am. Inst. Mining Met. Eng. Tech. Publ.* 2303 (1948).
4. ANISIMOV, S. M., and ZAPEVALOV, G. G., *Sb. Tr. Tsentral. Nauch.-Issledovatel. Lab. Zavoda "Elektrosink"* **1937**, 233–67. *C.A.* **33**, 7693 (1939).
5. FILIPPOVA, N. A., and SUDILOVSKAYA, E. M., *Analiz Rud Tsvetnykh Metal. i Produktov ikh Pererabotski* **1956**, No. 12, 14–23. *C.A.* **51**, 14485 (1957).
6. FILIPPOVA, N. A., KOROSTELEVA, V. A., and CHU, Y.-Y., *Zavod. Lab.* **27**, 1346–52 (1961). *C.A.* **56**, 9406 (1962).
7. IONESCU, M., and PAVEL, R., *Rev. minelor.* **9**, 39–44 (1958). *C.A.* **53**, 12092 (1959).
8. SUVOROVA, M. V., *Sbornik Nauch. Trudov. Vsesoyuz. Nauch.-Issledovatel. Gornomet. Inst. Tsvetnykh Metal.* **1960**, 442–8. *C.A.* **56**, 926 (1962).
9. KHRISTOFOROV, B. S., *Rudnyi Altai, Sovet. Narod. Khoz. Vostock.-Kazakhstan. Ekon. Admin. Raiona* **1958**, No. 3–4, 47–8. *C.A.* **54**, 18174 (1960).
10. SUDILOVSKAYA, E. M., *Analiz Rud Tsvetnykh Metal. i Produktov ikh Pererabotki, Sbornik Nauch. Trudov* **1958**, No. 14, 129–37. *C.A.* **53**, 13873 (1959).
11. KUDENKO, A. G., and PASHEVKINA, O. N., *Zavod. Lab.* **36**, 139–40 (1970). *C.A.* **73**, 31313 (1970).
12. FILIPPOVA, N. A., and KOROSTELEVA, V. A., *Zavod. Lab.* **25**, 535–9 (1959). *C.A.* **53**, 16805 (1959).
13. SHARYBKINA, M. A., FILIPPOVA, N. A., and SAMOKHVALOVA, L. G., *Zavod. Lab.* **37**, 1431–2 (1971). *C.A.* **76**, 80779 (1972).
14. BIRGER, N. I., ROZENBLYUM, E. N., and KOSTYCHOV, V. A., *Trudy Leningrad. Krasnoznamenskogo Khim.-Technol. Inst.* **1940**, No. 8, 150–3. *C.A.* **36**, 5725 (1942).
15. ISHII, R., and TANINO, K., *Bull. Inst. Phys. Chem. Research* **22**, 498–503 (1943). *C.A.* **42**, 8702 (1948).
16. MITYAEVA, P. A., and DEMCHENKO, R. S., *Tr. Inst. Met. Obogashch., Akad. Nauk Kaz. S.S.R.* **24**, 90–3 (1967). *C.A.* **68**, 26712 (1968).
17. RIKKERT, I. E., *Zavod. Lab.* **8**, 164–8 (1939). *C.A.* **33**, 9193 (1939).

18. Black, R. M., *Analyst* **76**, 208–11 (1951).
19. Zagorski, Z., *Chem. Anal.* (*Warsaw*) **1**, 188–98 (1956). *C.A.* **51**, 4200 (1957).
20. Ten'kovtsev, V. V., and Bagdasarov, K. N., *Zavod. Lab.* **22**, 657–8 (1956). *C.A.* **50**, 15341 (1956).
21. Filippova, N. A., Martynova, L. A., Selezneva, M. N., and Stepareva, V. N., *Sb. Nauch. Tr., Nauch.-Issled. Inst. Tsvet. Metal.* **34**, 179–82 (1971). *C.A.* **77**, 172271 (1972).

Magnesium

Phase analysis of magnesium compounds is confined to a differentiation of metallic magnesium from its oxide.

In one method, a 1–5 g sample is stirred 1–2 minutes with 20–50 ml of a 5% solution of chromium trioxide. It is filtered rapidly, and washed with water and 1% ammonium hydroxide until colourless. The filtrate contains the magnesium from magnesium oxide; metallic magnesium remains in the precipitate [1].

Another recommendation is to agitate for 5 minutes at room temperature a 0·5 g sample in 20 ml of 1N acetic acid containing 62 g/litre of potassium dichromate. Filter, wash with 1% ammonium hydroxide, dissolve the residue of metallic or "active" magnesium in sulphuric acid and determine by any appropriate method [2].

In a study of the determination of magnesium oxide in the presence of metallic magnesium, it was reported that under the optimum conditions of a contact time of 1 hour in 2% potassium dichromate, 94–95% of the magnesium oxide and 5–7% of the metallic magnesium was dissolved [3].

A different approach was used to differentiate magnesium oxide from metal in slags. Both the oxide and metal are leached from a magnesium fluoride slag with a buffered solution of EDTA, and total magnesium is determined by titration. In another sample, metallic magnesium is isolated by potassium dichromate and determined by titration. Magnesium oxide is found by difference [4].

To 1 g of slag add 25 ml of 0·1M EDTA and 10 ml of 6% ammonium chloride/75% ammonium hydroxide buffer. Boil for 20 minutes, cool, filter, adjust to *p*H 10 with ammonium hydroxide. To the filtrate add sulphuric acid, and evaporate to charring; add nitric acid to destroy organic matter, and evaporate to dryness. Dissolve the residue in water, add 10 ml of buffer solution, boil, filter, and adjust the filtrate to *p*H 10 with ammonium hydroxide. Add 2 ml of 10% potassium cyanide, 0·5 g of Eriochrome Black T, and titrate with 0·1M EDTA to obtain total magnesium.

To another 1-g sample add 15 ml of 10% potassium dichromate, 1 ml of a wetting agent, and allow to stand for 1 minute. Add 50 ml of acetic acid, stir for 20 minutes, filter, and wash. Macerate the paper and residue with 25 ml of 0·1M EDTA and 10 ml of buffer solution, boil and cool. Add buffer, filter, evaporate, oxidize the filtrate and analyse for magnesium as described in the preceding paragraph. This is metallic magnesium. Magnesium oxide is calculated from total magnesium − metallic magnesium = magnesium from magnesium oxide [4].

REFERENCES

1. CHUGUNOVA, V. I., and IVANOVA, A. P., *Zavod. Lab.* **13**, 1163–4 (1947). *C.A.* **43**, 4182 (1949).
2. POGRANICHNAYA, R. M., NERUBASHCHENKO, V. V., and GOUCHAROVA, V. P., *Zavod. Lab.* **37**, 537–8 (1971). *C.A.* **75**, 58327 (1971).
3. OSPANOV, K. K., and ALIMPEVA, S. D., *Zavod. Lab.* **37**, 1045–6 (1971). *C.A.* **76**, 30344 (1972).
4. MCKEND, J., *Anal. Chem.* **32**, 1193–6 (1960).

Manganese

A few investigations have been reported on the phase analysis of manganese ores and compounds.

One procedure for ores recommends treating the samples with ammonium hydroxide and ammonium chloride to dissolve manganous and ferrous compounds. Filter, treat the residue with a mixture of 2N sulphuric acid and hydrofluoric acid to determine trivalent manganese. The residue from the trivalent manganese solution is analysed for quadrivalent manganese [1].

Another study of natural and reduced manganese ores indicated that a 1-hour extraction at 70°C with 30 ml of methanol containing 5 g mercuric chloride dissolves metallic manganese but only negligible manganese monoxide. The latter may also be extracted from the residue, obtained after treatment by methanolic cupric chloride solution, by a solution of 6N ammonium sulphate at 100°C for 30 minutes. Trivalent and quadrivalent manganese oxides in ores were also examined [2].

Several workers have proposed methods for manganese oxides of various valencies when these are present together. The directions for one are as follows. Heat the sample with 20 ml of 6N ammonium sulphate for 20–25 minutes with stirring, filter, and wash. Manganese derived from MnO is in the filtrate. Treat the residue for 1 hour at 70–75°C with 15 ml of a solution containing 15 g metaphosphoric acid in 100 ml of sulphuric acid. Filter, wash with 25% sulphuric acid, and determine manganese in the filtrate; this represents the manganese derived from Mn_2O_3. Place a separate sample in a known volume of standard sodium oxalate solution, add 10 ml of 25% sulphuric acid, and keep at 70°C until dissolved. Filter, wash, and titrate the excess sodium oxalate with standard 0·1N potassium permanganate to obtain MnO_2 [3].

In another publication, it is shown that MnO_2 is practically insoluble in $K_2P_4O_7$ solution heated at 100°C for 45–50 minutes, and Mn_2O_3 and a part of Mn_3O_4 dissolves in hot 1:5 sulphuric acid containing 1–2 ml of normal oxalic acid [4].

Another determination of manganese oxides of various valencies has been described [5]. To the sample add 50 ml of 3M ammonium sulphate solution and heat, with stirring, for 1 hour. Filter, wash, and determine manganese in the filtrate. If alkaline earths are present, carry out the solution and filtration in an inert atmosphere. This corresponds to MnO, but includes a small part of Mn_2O_3 and a large amount of manganese from $MnCO_3$. Determine MnO_2 in the residue. The difference between the total MnO_2 determined directly, and the MnO_2 in the residue corresponds to the Mn_2O_3 decomposed by ammonium sulphate.

To determine total MnO_2, add to the sample 0·5 g potassium iodide, 10 ml of 0·1M EDTA, 0·5 g sodium acetate, 10 ml of 5% acetic acid, and mix in a closed flask for 7 minutes. Dilute with water and titrate with standard sodium thiosulphate. To determine Mn_2O_3, dissolve the sample by heating on a water bath for 1 hour with 60 ml of 1% nitric acid. Filter, wash, and determine manganese in the filtrate. The total amount of manganese found corresponds to that from the MnO and half of that from Mn_2O_3. Transfer the filter with the residue to a flask, and determine MnO_2 as described above. The difference between the present amount and that determined directly by the above procedure corresponds to the second half of the manganese from Mn_2O_3 [5].

The determination of Mn_2O_3 can also be carried out by heating the sample on a water bath for 1 hour in a mixture of 100 ml of 10% sodium pyrophosphate and 2 ml of 1:1 sulphuric acid, filtering, adding a known amount of ferrous ammonium sulphate and titrating the excess with standard potassium permanganate. To determine Mn_3O_4, decompose the sample with nitric acid to liberate Mn^{+4} from MnO, Mn_2O_3, and Mn_3O_4; determine MnO and Mn_2O_3 as described above, and obtain Mn_3O_4 by difference [5].

A phase analysis of manganese in carbon steel and low alloy steel has been reported. Manganese as carbide, oxide, sulphide, and solid solution is separated from the residue and electrolyte by various solvents and techniques. The manganese compound is isolated from steel by making steel the anode and dissolving in a mixture of 1% sodium chloride/5% EDTA, at *p*H 6–7, with a current density of 50 mA/cm^2 for 2 hours. The steel sample is covered with a close texture filter paper as a diaphragm [6].

REFERENCES

1. ZAN'KO, A. M., and STEFANOVSKII, V. F., *J. Applied Chem.* (U.S.S.R.) **9**, 2192–2202 (1936). *C.A.* **31**, 5717 (1937).
2. WATANABE, S., *Nippon Kinzoku Gakhaishi* **24**, 217–25, 346–50, 401–13 (1960). *C.A.* **60**, 15130 (1964).
3. LAVKRUKHINA, A. K., *Zhur. Anal. Khim.* **4**, 40–5 (1949). *C.A.* **44**, 481 (1950).
4. FIKHTENGOT'TS, V. S., *Zavod. Lab.* **21**, 1036–8 (1955). *C.A.* **50**, 9216 (1956).
5. BAKEAN, J., *Hutnicke listy* **14**, 1084–6 (1959). *C.A.* **58**, 7371 (1963).
6. WAKAMATSU, S., *Tetsu To Hagane* **58**, 1485–94 (1972). *C.A.* **77**, 134768 (1972).

Molybdenum

A differentiation of sulphide molybdenum from oxidized forms is important in mine and concentrator samples, because only the sulphide form is recoverable by conventional flotation practice. In fact, some producers determine sulphide molybdenum rather than total molybdenum.

To 1–5 g of −100 mesh sample add 50 ml of 30% hydrochloric acid, and boil gently for 20 minutes. Filter through Whatman No. 40 paper with pulp, and wash thoroughly with hot water. The filtrate contains the oxide molybdenum, which may be analysed by any appropriate method. Molybdenite, or molybdenum disulphide, is not dissolved and remains in the residue on the paper. Digest the latter with nitric and sulphuric acids, finally adding a little perchloric acid, if necessary, to destroy the organic matter; determine this sulphide molybdenum by any convenient procedure [1].

Several workers have reported phase analyses of various molybdenum compounds. In molybdenum powder, molybdenum trioxide has been determined by heating with pyridine on a water bath, and extracting with 12% ammonium hydroxide; no oxidation of molybdenum to trioxide occurs [2].

In another investigation, molybdenum dioxide was determined by the quantity consumed in reducing silver from silver nitrate in the presence of ammonium hydroxide. Molybdenum disulphide is found by solution of the latter and of molybdenum dioxide in boiling 20% sodium hydroxide solution; with the content of molybdenum dioxide known, the disulphide is obtained by difference. Calcium molybdate is dissolved in boiling 5% glycol, lead molybdate in boiling 15–20% sodium hydroxide, and iron molybdate in 3% sodium citrate [3].

A phase analysis of molybdenum-containing precipitates from molybdate solutions reduced with hydrogen has been published. The 0·1-g sample is extracted with 30 ml of ferrous ammonium sulphate for 24 hours, 5 ml phosphoric acid is added, and the solution is titrated with 0·1N potassium permanganate [4].

In a mixture of molybdenum trioxide and disulphide, the former can be removed by leaching the dried, oil-free sample with hot 50%

ammonium hydroxide. The filtrate contains molybdenum trioxide, the disulphide remaining in the residue. Total molybdenum can be determined in a separate sample by roasting in a platinum crucible for 45 minutes at 575°C to oxidize all molybdenum to the trioxide form [5].

The phase analysis of molybdenum in low-alloy steel has been reported. Cover the sample with a filter paper diaphragm, connect as an anode in 100–300 ml of 1% sodium chloride/5% EDTA solution at *p*H 6–7, and a current density of 50 mA/cm² for 1–2 hours. To the residue add 20 ml of 6N hydrochloric acid, flush the mixture with argon for 10 minutes, and filter. To the filtrate add nitric and perchloric acids, and evaporate to fumes. Determine in this portion the molybdenum derived from $(Fe,Mo)_3C$. Digest the paper and residue from the filtration and determine the molybdenum from Mo_2C. The molybdenum in the original electrolyte is that present in solid solution [6].

REFERENCES

1. YOUNG, R. S., *Chemical Analysis in Extractive Metallurgy*, London, Charles Griffin, 1971.
2. PAVELKA, F., LAGHI, A., and ZUCCHELLI, A., *Mikrochemie verein. Mikrochim. Acta* **31**, 97–101 (1943). *C.A.* **39**, 2710 (1945).
3. ZABLOTSKAYA, V. L., *Trudy Severo-Kavkas. Gorno-Met. Inst.* **1956**, No. 13, 113–20. *C.A.* **52**, 15331 (1958).
4. SOLNTSEV, N. I., CHUDINA, R. I., SAVINA, E. V., and KULICHIKHINA, R. D., *Sb. Nauchn. Tr., Gos. Nauchn.-Issled. Inst. Tsvetn. Metal.* No. **18**, 155–64 (1961). *C.A.* **60**, 2321 (1964).
5. ANGELOTTI, N. C., and GOOCH, E. G., *Anal. Chim. Acta* **58**, 445–7 (1972).
6. WAKAMATSU, S., *Tetsu To Hagane* **58**, 472–81 (1972). *C.A.* **76**, 148476 (1972).

Nickel

A distinction between metallic nickel, oxide nickel, and sulphide nickel is important in many processes of primary extraction, powder metallurgy, catalyst production, and other industrial activities. A number of publications, consequently, have dealt with phase analyses of nickel ores and compounds.

1. Metallic and Sulphide Nickel

A. Chlorine–Methanol

In the nickel industry, metallic and sulphide nickel have been frequently differentiated from the oxide form of the element by leaching with a chlorine–methanol solution. Metallic and sulphide nickel dissolve, whereas the oxide remains insoluble. As with all phase analyses, the method is empirical and the result is only approximate. For process control in such fields of extractive metallurgy as the roasting or reducing of ores, concentrates, mattes, and other materials, it can provide very useful results. The procedure below is typical, but of course may be modified to suit the individual product.

Weigh into a dry, tall 400-ml beaker a suitable quantity of –200 mesh material, depending on the nickel content and the analytical method to be employed. Add at least ten times the sample weight of anhydrous methanol, stir and place in a fume cupboard. Pass a vigorous stream of dry gaseous chlorine from a cylinder into the dilute pulp for 10 minutes. Nickel in metallic form, and nearly all the sulphide nickel, will dissolve in the chlorine–methanol solution; nickel oxides are virtually unattacked. Filter, wash the residue thoroughly with anhydrous methanol, and evaporate the filtrate to dryness. Moisten with nitric acid and evaporate, repeating if necessary, to destroy organic matter. Add water, boil, cool, make to a definite volume, and determine the metallic and sulphide nickel in a suitable aliquot by any convenient method. The oxide nickel in the residue after filtration may be determined by destroying the filter paper with nitric and sulphuric acids, and carrying out any appropriate procedure for nickel [1].

B. Bromine–Methanol

The use of a bromine–methanol solution to selectively dissolve metallic nickel from a mixture with nickel oxide has been described. To 0·5 g of −100 mesh dry sample in an Erlenmeyer flask add 50 ml of anhydrous methanol containing 5% bromine. Boil under a reflux condenser for 15 minutes, cool, filter, and wash with about 50 ml of anhydrous methanol. To the filtrate add 20 ml of 1:1 hydrochloric acid, 5 ml of 75% hydroxylamine hydrochloride solution, and mix. Allow to stand until reduction of bromine is complete. Add 3% hydrogen peroxide, boil for 5 minutes to decompose the excess of this reagent, cool, dilute to 200 ml, and determine nickel by atomic absorption. Solution of the oxide phase does not occur if free and combined water is removed from the sample [2].

In the application of the bromine–methanol method to reduced garnierite ores, it was found that the oxides of magnesium and nickel are hardly attacked by bromine and are nearly insoluble in anhydrous methanol. This procedure has an advantage over the mercuric chloride extraction in the presence of magnesium oxide; the latter will precipitate a basic mercuric chloride [3].

A novel application of the methanol–bromine method for metallic nickel has been reported. At −20°C methanol–bromine readily dissolves the metallic nickel binder from sintered titanium carbide. The low temperature, obtained by alcohol and dry ice, eliminates the need for anhydrous reagents and inert atmospheres [4].

Metallic nickel in catalysts has been found by solution in bromine–methanol in an evacuated glass assembly [5].

It has been reported that, in roasted and reduced nickel ores, metallic nickel is dissolved quantitatively when a −180 mesh sample is treated for 15 minutes at 30°C with anhydrous methanol containing 4% iodine [6].

2. Metallic Nickel

A. Mercuric Chloride

The determination of metallic nickel by its interaction with a mercuric chloride solution has been published. In one investigation, two procedures are proposed. In a 300-ml flask under a condenser, boil 100 ml of a solution containing 18·5 g/litre of mercuric chloride. When the solution boils, add a sample containing not more than 0·1 g of metallic nickel. Continue boiling for 2 hours, allow to cool, filter, and determine nickel in the filtrate. Nickel oxide remains in the residue.

For the second procedure, the same practice is followed, but the solution consists of 100 ml of 18·5 g/litre mercuric chloride, and 25 ml of 80 g/litre potassium cyanide brought to *p*H 9 with additions of 6N hydrochloric acid. Boiling is continued for 6 hours. This second procedure gives slightly better results, but the time is three times as long [7].

For a determination of nickel metal and oxides in catalysts, the sample was boiled for 20 minutes with saturated mercuric chloride. After cooling, ammonium hydroxide was added to precipitate mercury as $(NH_2Hg)Cl$, and the solution was filtered. Metallic nickel is in the filtrate and nickel oxides remain in the residue; both can be analysed for nickel by any suitable method [8].

B. HYDROGEN EVOLUTION

The volume of hydrogen evolved when metallic nickel is treated with acid furnishes another approach to nickel phase analysis. Publications have appeared on the volumetric determination of hydrogen liberated, by dissolving in dilute hydrochloric acid the metallic nickel in a mixed catalyst [5], by treating nickel and its oxide in 4N hydrochloric acid in a flask resembling a Van Slyke apparatus [11], and by reacting nickel and nickel sulphide with 1:1 hydrochloric acid and removing hydrogen sulphide and hydrogen chloride [12].

3. Nickel Sulphide

A useful procedure for nickel sulphide in the presence of metallic nickel and nickel oxide has been developed for ores and metallurgical products. It depends on the solubility of nickel sulphide in a mixed solution of ammonium citrate and strong hydrogen peroxide. Metallic nickel and the oxide are not attacked if the contact time does not exceed 2 hours at room temperature [9, 10].

Transfer 0·2–2 g of –150 mesh sample to a 100-ml Kohlrausch flask. Add 50 ml of a solution containing 75·5 g of di-ammonium hydrogen citrate in 1 litre of water, and 25 ml of 100 volumes hydrogen peroxide. Stopper the flask and agitate for 2 hours at room temperature. Dilute to volume, mix, allow to settle, filter if necessary and determine nickel by atomic absorption or any suitable method. Nickel in the filtrate is that derived from nickel sulphide; nickel in the residue is metallic and oxide nickel. If nickel sulphate is present in the original sample it will report with the sulphide phase [9, 10]. Nickel sulphate, however, can be removed initially by a simple water leach.

4. Miscellaneous Separations

A satisfactory separation of nickel oxide, sulphide, and silicate has been reported. When the mixture was heated at 60°C for 6 hours with a solution containing 80 ml of acetic acid, 80 ml water, and 40 ml of 3% hydrogen peroxide, only the nickel sulphide was dissolved. Boiling the residue with 10% sulphuric acid for 10 minutes dissolved nearly all the nickel silicate, leaving the oxide in a final residue [13].

A determination of metallic nickel in the presence of nickel oxide has been published, in which the 0·05-g sample is stirred for 1–2 hours at 60°C with 50 ml of a solution containing 60 g of ferrous ammonium sulphate and 1 ml of 0·001N sulphuric acid. The reaction is carried out in an atmosphere of nitrogen, and final titration of an aliquot is made with potassium dichromate, using sodium diphenylamine sulphonate as indicator [14].

In foundry products containing metallic and oxide nickel, the former could be selectively dissolved in a solution of silver thiocyanate/ammonium thiocyanate if the sample particles were finer than −230 mesh [15].

REFERENCES

1. YOUNG, R. S., *Chemical Analysis in Extractive Metallurgy*, London, Charles Griffin, 1971.
2. KINSON, K., DICKESON, J. E., and BELCHER, C. B., *Anal. Chim. Acta* **41**, 107–12 (1968).
3. SHIRANE, Y., *Kyushu Kozan Gakkai-Shi* **37**, 127–33 (1969). *C.A.* **71**, 131329 (1969).
4. VIOLANTE, E. J., *Anal. Chem.* **33**, 1600–2 (1961).
5. URBAIN, H., BACAUD, R., CHARCOSSET, H., and TOURNAYAN, L., *Chim. Anal.* **50**, 242–6 (1968). *C.A.* **69**, 64365 (1968).
6. MAEDA, F., TAKEYAMA, S., and GOTO, H., *Bunseki Kagaku* **16**, 54–5 (1967). *C.A.* **67**, 122009 (1967).
7. EK, C., *Anal. Chim. Acta* **14**, 311–17 (1956).
8. MISHKAREVA, L. V., and STERLIN, B. Y., *Maslob.-Zhir. Prom.* **30**, (9), 14–15 (1964). *C.A.* **66**, 25805 (1967).
9. KATSNEL'SON, E. M., and OSIPOVA, E. Y., *Obogashchemie Rud* **5**, No. 4, 24–6 (1960). *C.A.* **55**, 26854 (1961).
10. BARKER, C. W., personal communication, 1971.
11. KREJCAR, E., *Chem. listy* **52**, 2410–11 (1958). *C.A.* **53**, 8935 (1959).
12. ZAVGORODNYAYAYA, E. F., MANGUSHLO, L. N., and MIRONOVA, V. M., *Zavod. Lab.* **33**, 949 (1967). *C.A.* **67**, 113431 (1967).
13. SHAKHOV, G. A., and VOSKRESENSKAYA, M. M., *Zavod. Lab.* **13**, 156–60 (1947). *C.A.* **42**, 485 (1948).
14. KREJCAR, E., *Chem. listy* **51**, 2137 (1957). *C.A.* **52**, 2645 (1958).
15. KRAFT, G., *Z. Anal. Chem.* **209**, 150–6 (1965). *C.A.* **62**, 15418 (1965).

Niobium

The separation of metallic niobium, niobium carbide, and niobium pentoxide has been reported. To 0·2 g of sample add 5 ml of water, 5 ml nitric acid, and 3–4 ml of hydrofluoric acid. Heat, cool, add 20 ml of 1:4 sulphuric acid, filter, and evaporate. Determine niobium on this filtrate; it represents metallic niobium and niobium carbide. The residue on the filter paper contains niobium pentoxide, which may be analysed by any convenient procedure. To another 0·2-g sample add 60 ml 2:1 hydrofluoric acid, heat for 2 hours, filter, and evaporate. Determine niobium in this filtrate; it represents metallic niobium and niobium pentoxide. From these two determinations, the three phases can be readily calculated. It is stated that, for a content of single components ranging from 15–50%, the error is only 0·8 to 1·8% [1].

REFERENCE

1. Kozyreva, L. S., Kuteinikov, A. F., and Romantovskaya, N. I., *Zavod. Lab.* **33**, 295–6 (1967). *C.A.* **67**, 39861 (1967).

Nitrogen

Nitrates and nitrites may occur together in waters. If the phenol-disulphonic acid method is used for nitrates, nitrites respond like nitrate. It is necessary to subtract the value for nitrite nitrogen, determined in a separate sample, by the formation of a reddish-purple azo dye produced at *p*H 2·0–2·5 by the coupling of diazotized sulphanilic acid with naphthylamine hydrochloride. If the brucine method is used to determine nitrates, no interference is encountered from nitrite [1, 2].

Nitrate and nitrite in mixtures have been determined with a nitrate ion electrode. The nitrate content is first determined, then the nitrite is oxidized with potassium permanganate in acid solution, and a second measurement gives the sum of nitrate + nitrite [3].

REFERENCES

1. American Public Health Association, *Standard Methods for the Examination of Water and Wastewater*, New York, American Public Health Association, 1965.
2. A.O.A.C., *Official Methods of Analysis of the Association of Official Analytical Chemists*, Washington, D.C., Association of Official Analytical Chemists, 1965.
3. Morie, G. P., Ledford, C. J., and Glover, C. A., *Anal. Chim. Acta* **60**, 397–403 (1972).

Phosphorus

With rare exceptions, phosphorus occurs in nature, in both organic and inorganic forms, as the orthophosphate. In phosphorus compounds produced artificially, however, there may be chain phosphates or polyphosphates, and ring phosphates or metaphosphates. Orthophosphates form yellow precipitates with ammonium molybdate in cold acid solution, but neither the poly-(chain) nor meta-(ring) phosphates show this behaviour.

1. Polyphosphate in the presence of Orthophosphate

In water samples, polyphosphate can be determined in the presence of orthophosphate by the following procedure. In one 100-ml sample, determine phosphorus by the reaction of orthophosphate with ammonium molybdate in an acid medium to form a heteropoly acid, molybdophosphoric acid, which is reduced to the intensely coloured complex, molybdenum blue, by the combination of amino naphthol sulphonic acid and sulphite reducing agents. Alternatively, substitute stannous chloride for amino naphthol sulphonic acid as the reducing agent. Polyphosphates do not interfere.

To a second 100-ml sample add a drop of phenolphthalein indicator solution and then an acid solution containing 300 ml sulphuric acid and 4 ml nitric acid per litre until the colour is discharged; finally add 1 ml excess acid solution. Boil for at least 90 minutes, adding water to keep the volume between 25 and 50 ml. Polyphosphates are hydrolyzed by this treatment to orthophosphate. After cooling, determine phosphorus by the method outlined above. This gives total phosphate, and by subtracting from this the orthophosphate found in the first sample, the polyphosphate is obtained [1].

2. White Phosphorus, Phosphorous Acid, and Phosphoric Acid

Directions have been given for an analysis of a mixture of white phosphorus, phosphorous acid, and phosphoric acid [2]. An aqueous solution of this mixture is obtained when a sample containing phosphorus, phosphorus trichloride, and phosphorus oxychloride is treated with water for 15 minutes. The procedure was developed for the analysis of 3–5 g of a mixture of P, PCl_3, and $POCl_3$ which was hydrolyzed for 15 minutes with 200 ml of water; red phosphorus would remain as an insoluble residue.

Extract the aqueous solution of about 200-ml volume in an Erlenmeyer flask by shaking for 1 minute with 75 ml of benzene. Transfer to a 50-ml separatory funnel, wash the flask twice with 20 ml of water and 20 ml of benzene, and add the washings to the funnel. Shake for two minutes, allow to separate, and transfer the aqueous phase into a second separatory funnel. Add 50 ml of benzene to the aqueous phase, shake for 2 minutes, draw the aqueous layer into a third funnel, and again wash. Transfer the aqueous phase into a 500-ml volumetric flask. Collect all the benzene phases in the first funnel, and wash both the second and third funnels once with 20 ml of water and 20 ml of benzene. Combine these washings with the contents of the first funnel. Shake for 1 minute and add the aqueous layer to the volumetric flask. Wash the benzene by shaking twice with 20 ml of water, add the water to the volumetric flask, and make up to volume.

Transfer the benzene–phosphorus solution to 50 ml of 25% copper nitrate solution in a 500-ml Erlenmeyer flask provided with a ground-glass stopper. If more than 40 mg of phosphorus was extracted, make the benzene solution to volume and transfer an aliquot to the copper nitrate solution. Wash the funnel twice with 10 ml of benzene and add these washings to the flask containing the copper nitrate. Stopper the flask, and shake at intervals for 15 minutes. Boil until all benzene is removed. Add carefully 25 ml of nitric acid, and, when the reaction has ceased, 25 ml of hydrochloric acid. Boil until nitrous fumes are removed, cool, and determine phosphorus in the solution or in an aliquot. This represents white phosphorus.

Transfer an aliquot of the aqueous solution, obtained in the original extraction procedure, to a mixture of 50 ml of 0·1N iodine solution and 50 ml of a saturated solution of sodium bicarbonate. Allow to react in the dark for 30 minutes, and titrate the excess of iodine with standard arsenious solution. This represents the phosphorus derived from phosphorous acid. One ml of 0·1N iodine solution = 1·549 mg of trivalent phosphorus.

Transfer another aliquot of the aqueous solution to a 600-ml beaker. Add 15 ml of nitric acid and boil for 5 minutes. Add 10 ml of perchloric acid, heat to fumes of the latter, and continue fuming for 5 minutes. Cool and determine phosphorus by the standard ammonium molybdate procedure. This gives the sum of phosphorus from phosphoric and phosphorous acids; subtract the value of the latter, obtained in the previous paragraph, to give the phosphorus derived from phosphoric acid.

3. Orthophosphorous Acid, Hypophosphorous Acid, Hypophosphoric Acid, and Orthophosphoric Acid

A liquid mixture of orthophosphorous acid, hypophosphorous acid, hypophosphoric acid, and orthophosphoric acid has been analyzed, as has a mixture of ortho-, pyro-, and tri-phosphate [2].

Transfer an aliquot of the neutral solution to a 500-ml flask provided with a ground-glass stopper. Add 50 ml of 0·2M sodium bicarbonate solution which has been previously saturated with carbon dioxide. Add an excess of 0·1N iodine solution, stopper the flask, and allow to react for 45 minutes. Titrate the excess of iodine with a standard solution of arsenite or sodium thiosulphate. One ml of 0·1N iodine solution = 4·10 mg of orthophosphorous acid, H_3PO_3.

Transfer another aliquot of the neutral solution into a flask provided with a ground-glass stopper. Add 10 ml of 15% sulphuric acid, and an excess of 0·1N iodine solution. Stopper the flask and allow to react for 10 hours. Add a pasty suspension of sodium bicarbonate slowly until evolution of carbon dioxide ceases. Add 50 ml of 0·2M sodium bicarbonate which has been saturated with carbon dioxide. Stopper the flask, allow to stand for 1 hour, and titrate the excess of iodine with standard sodium thiosulphate. This gives a value corresponding to the sum of orthophosphorous acid originally present plus the phosphorous acid formed by oxidation of the hypophosphorous acid in acid solution. The iodine consumption due to phosphorous acid originally present must be determined, as in the previous paragraph, and has to be deducted from the volume of 0·1N iodine solution used. After this deduction, 1 ml of 0·1N iodine solution = 1·65 mg of hypophosphorous acid, H_3PO_2.

To another aliquot of the solution that has been neutralized against phenolphthalein, add 2 drops of 10% acetic acid, 20–25 ml of 20% sodium acetate solution, and 25 ml of a saturated barium nitrate solution. Separate the precipitate by centrifuging 20–30 minutes. Stir the precipitate with water, and again centrifuge; discard the solutions. Dissolve the precipitate, which is free of hypophosphite and consists of the barium salts of orthophosphorous, hypophosphoric, and orthophosphoric acids, in 20–25 ml of 10% phosphoric acid. Add a few ml of ether and precipitate silver hypophosphate by adding a 100% excess of 0·1N silver nitrate. Mix well by shaking, and filter immediately under slight suction into a 250-ml ground-glass-stoppered flask. Wash several times by decantation and finally several times on the filter. Acidify the filtrate without delay and titrate the excess of silver with ammonium thiocyanate. One ml of 0·1N silver nitrate corresponds to 8·099 mg of hypophosphoric acid, $H_4P_2O_6$.

Transfer an aliquot of the original neutral solution to a beaker, add nitric and hydrochloric acids, and evaporate to fumes. Repeat this operation with more nitric and hydrochloric acids. Determine phosphorus by the standard ammonium molybdate method, and deduct from this value for total phosphorus the amounts of phosphorus found in the lower oxidation states of orthophosphorous acid, hypophosphorous acid and hypophosphoric acid as described in the preceding paragraphs. The net result represents orthophosphoric acid, H_3PO_4.

4. White and Red Phosphorus

White phosphorus can be extracted from a mixture of white and red phosphorus by carbon disulphide. Red phosphorus is nearly insoluble in any solvent; the white form is soluble in carbon disulphide and slightly soluble in other solvents such as ether, carbon tetrachloride, benzene, and a mixture of carbon disulphide, ether, and alcohol [3].

REFERENCES

1. AMERICAN PUBLIC HEALTH ASSOCIATION, *Standard Methods for the Examination of Water and Wastewater*, New York, American Public Health Association, 1965.
2. FURMAN, N. H., ed., *Scott's Standard Methods of Chemical Analysis*, 6th ed., Vol. 1, Princeton, N. J., Van Nostrand, 1962.
3. HALMANN, M., ed., *Analytical Chemistry of Phosphorus Compounds*, New York, Wiley-Interscience, 1972.

Rhenium

A chemical phase analysis of the rhenium compounds formed during the oxidation of rhenium disulphide has been described [1]. Agitate a 0·5 g sample for 10 minutes with 100 ml of water at room temperature, filter, and wash thoroughly. Rhenium heptoxide, R_2O_7, dissolves and may be determined in the filtrate.

Heat the residue with 60–70 ml of hydrochloric acid on a water bath, filter, and wash. Rhenium dioxide is soluble and passes into the filtrate. Leach the second residue with 60–70 ml of 10% ferric sulphate containing 25 ml of sulphuric acid per 100 ml, filter, and wash. Rhenium trioxide dissolves and is determined in the filtrate, after addition of 5 ml phosphoric acid, by titration with 0·1N potassium permanganate.

Treat the third residue with 60 ml of a solution of 3 parts methanol to 1 part water, containing 4% bromine, on a water bath for two hours at 60–70°C. Filter and wash with methanol; metallic rhenium is in the filtrate and rhenium disulphide remains in the final residue. Excess bromine is removed by boiling on a water bath, the residue sample is evaporated three times with hydrochloric acid, diluted, and rhenium determined in the solution.

REFERENCE

1. Babenko, N. L., and Sukhorukova, N. V., *Zavod. Lab.* **35**, 1047–51 (1969). *C.A.* **72**, 38570 (1970).

Selenium

A phase analysis of selenium compounds found in lead smelter dusts has been reported. Treat the sample at room temperature with a mixture of anhydrous methanol and anhydrous sodium sulphate, filter, and wash with anhydrous methanol. Selenium dioxide passes into the filtrate.

Weigh another portion of the dust and boil with 0·25M Trilon solution at *p*H 8·5 to extract the selenium of the selenites and selenium dioxide. After filtration, boil the residue with 1·5M sodium sulphite, filter, and wash. Elemental selenium passes into the filtrate. Boil the second residue with 7N nitric acid to extract the selenium of zinc and lead selenides. After filtration and washing, the final residue contains the selenium of mercuric selenide [1].

If the sample when received is in the liquid state, as in leach plants or electrolytic refineries, selenium (IV) can be separated from selenium (VI) by thin layer chromatography on silica gel with 1 butanol : 3 ammonium chloride [2].

REFERENCES

1. Filippova, N. A., Martynova, L. A., Savina, E. V., and Kulichikhina, R. D., *Zavod. Lab.* **26**, 401–10 (1960). *C.A.* **54**, 15071 (1960).
2. Mekra, H. P., Mittal, I. P., and Johri, K. N., *Chromatographia* **4**, 532–4 (1971). *C.A.* **76**, 94122 (1972).

Silicon

1. "Free Silica" or Quartz

Phase analysis of "free silica" or quartz received a big impetus many years ago when it was found that quartz particles in dust were more detrimental to health than silicate or other rock particles. The hazard of dusts in mining, quarrying, construction, and similar activities has resulted in the publication of many procedures for the determination of free silica.

A. Acid Treatment

(*1*) *Hydrofluosilicic Acid and Other Reagents*. One of the first methods to be developed consists of treatment of the dust of pulverized rock successively with hydrochloric, hydrofluosilicic, and hydrofluoric acids. Carbonates are decomposed by hydrochloric acid, and silicates by hydrofluosilicic acid; free silica is virtually unattacked and remains in the residue to be dissolved by hydrofluoric acid and determined by loss of weight on volatilization. The procedure outlined below is typical.

Place a 0·5 g sample of dust or −150 mesh rock in a platinum dish. If it contains organic matter, ignite for 30 minutes at 600–700°C. Cool, add hydrochloric acid, and warm until evolution of carbon dioxide ceases. Cool, dilute with water, filter through ashless filter paper, and wash thoroughly. Return the paper and precipitate to the platinum dish. Dry, ignite at 1000°C, cool, and weigh. Add hydrofluosilicic acid, cover the dish with a plastic or platinum cover, and allow it to stand at 20°C or below for 1–3 days, depending on the anticipated silicate content. If the latter is not known, run three samples and allow to digest for 1, 2, and 3 days, respectively.

Dilute with water, filter through ashless filter paper, and wash thoroughly. Return the paper and residue to the platinum dish, dry, ignite at 1000°C, cool, and weigh. The loss in weight due to the hydrofluosilicic acid treatment is silicate. To the residue in the platinum dish, add 3 ml of hydrofluoric acid, and place on the hot plate until all the acid has volatilized. Ignite at 1000°C, cool and weigh. Repeat this treatment, if necessary, until constant weight is obtained. The loss in weight due to the hydrofluoric acid treatment is free silica.

Hydrofluosilicic acid has a slow solvent action on free silica, and the results may be corrected by adding 0·7% per day, for each day of treatment with hydrofluosilicic acid, to the weight of free silica found [1, 2, 3, 4].

In another published procedure for quartz in the presence of silicates, a different mixture of acids is used. By dissolving 32 g of boric acid in 75 ml of 48% hydrofluoric acid, a solution of fluoboric acid is obtained which will give no test for fluoride. If 0·2 g of a mixture of silicate and silica is treated with 5 ml of fluoboric acid, 1 ml of phosphoric acid, and 2 ml of 2N ferric chloride solution, most silicates will be decomposed by heating 48 hours at 50°C, leaving quartz unattacked. If the yellow colour of ferric chloride fades, more of the latter should be added. The correction factor for quartz dissolved by this acid mixture is 0·34% per day [5, 6].

Free silica has been determined in iron ores by the following method. Treat the —120 mesh sample with a mixture of acetic acid and sodium acetate, of *p*H 2·5, at room temperature. Filter through a tared alumina crucible, wash thoroughly, and boil the residue in 20% sodium sulphide solution. Filter through the same crucible, wash thoroughly, transfer the residue to the original beaker, and boil with 100 ml of sodium sulphide solution. Filter through the same crucible, wash and transfer to the original beaker. Treat with 200 ml hydrochloric or nitric acid, filter through the crucible, wash, dry, and weigh. This represents free silica [7].

Free silica in coal dust has been found by the following procedure. Carefully ignite the sample in a muffle until all carbonaceous matter is destroyed. Treat the ash with hydrochloric acid until action ceases, dilute, boil, filter, and wash. Boil the residue with 25% sodium sulphide, cool, dilute, filter and wash. This residue contains silica, metal sulphides, and sesquioxides; remove the latter two with a mixture of 6N hydrochloric and 6N nitric acids, filter, wash and weigh the residue of free silica. Obtain a true value of the latter by the loss in weight through the classical evaporation with hydrofluoric and sulphuric acids [8].

(2) *Phosphoric Acid.* An important procedure for free silica is based on the fact that at 230–250°C phosphoric acid is converted to pyrophosphoric acid, which dissolves all silicates but not quartz. This is currently the most popular phase analysis for free silica. The method outlined below is typical.

Ignite a 0·5–2 g sample at 600°C for 1 hour to destroy organic matter, cool, treat with 35% hydrochloric acid on a water bath to decompose carbonates. Dilute, filter, wash, and treat the residue with 25 ml of phosphoric acid at 250°C for 30–60 minutes. Cool, dilute, filter, and wash with hot water. Add to the residue hot 5% sodium hydroxide to dissolve any freshly precipitated amorphous silica resulting from decomposition of silicates. Wash the residue thoroughly with hot water, 10% hydrochloric acid, and again with hot water. Carefully ignite at 1000°C, cool, weigh, add hydrofluoric and sulphuric acids, evaporate to dryness, ignite, and weigh. The loss in weight represents the free silica [9, 10, 11, 12, 13, 14, 15, 16, 17, 18, 19, 20].

A slightly different method is detailed below [9, 17]. Place 0·5 g of —200 mesh sample in a small beaker, add 25 ml of phosphoric acid, cover with a funnel, and place on an electric heater which has been pre-heated to provide a temperature about 250°C. When boiling subsides, swirl the beaker for 3 seconds at 1-minute intervals to minimize superheating. At the end of a boiling period of 12 minutes, remove, swirl, and cautiously remove the funnel so that adhering liquid runs down the side of the beaker. Allow to cool to room temperature, wash down the sides of the beaker with 125 ml of water at 60–70°C, and swirl to dissolve the syrupy phosphoric acid. Wash down the upper part of the beaker with 25 ml of water, and allow to stand for 1 hour. Filter on fine paper, using pulp; wash with cold and then hot 10% hydrochloric acid, and finally with 1–2 water washes. Ignite the paper carefully in a tared platinum crucible and heat at 950°C to constant weight. Moisten with 1:1 sulphuric acid and 5 ml or more of hydrofluoric acid. Heat, repeat the acid treatment to ensure complete volatilization of silica, ignite at 950°C, cool, and weigh. The residue is quartz or free silica.

Later refinements in a publication include a rotator during digestion, a voltage-controlled heater, a digestion of orthoclase for varying times, correction curves for the loss of quartz, and identification of undissolved minerals by petrographic means [17]. Quartz recovery from dust particles below 5 microns was 95%, with the digestion time reduced to 8 minutes.

Good agreement is generally obtained between phase analysis by the pyrophosphoric acid method and X-ray diffraction.

Pyrophosphoric acid reacts with most silicates in about 15 minutes at 250°C to form water-soluble compounds, but beryl, topaz, tourmaline, and zircon are more resistant than other common silicates.

Various modifications of the outlined method given above have been suggested. After the reaction with pyrophosphoric acid, an addition of fluoboric acid and tartaric acid may be made before filtration of silica [18, 20].

B. Fusion Treatment

In place of treatment with various acid solutions to isolate free silica, fusion of the sample with certain fluxes to achieve the same end has been recommended.

Ignition of the sample to decrease the solubility of quartz in alkaline solution, followed by fusion with potassium pyrosulphate at a low red heat, has been proposed; the free silica is not attacked. The silicic acid liberated by the decomposition of the silicates is extracted with sodium hydroxide solution under controlled conditions [21].

A double fusion of silicates with potassium bisulphate, followed by digestion in warm water, filtration, washing once with dilute hydrochloric acid, and finally with water, has been found to leave all quartz on the filter [22].

A rapid method for free silica in clay, kaolin, and bauxite consists of fusion with potassium bisulphate or pyrosulphate for 3–4 minutes at 800°C, digestion in 1 hydrochloric acid : 2 water, and filtration. Free silica remains in the residue [23].

Another investigator reports that quartz and opal are practically insoluble in fused sodium bisulphate [24].

A different flux is suggested in other publications. Equal parts of potassium bicarbonate and potassium chloride, or for a higher melting temperature, sodium bicarbonate and sodium chloride, will dissolve free silica, leaving the silicates practically untouched [25, 26]. The method is recommended for high silica contents; for less than 10% free silica the pyrophosphate procedure, leaving quartz unattacked, is preferred [26].

Other workers have proposed another flux. A sample of 0·1–0·2 g is fused with 1·7 g of flux, consisting of 1 g potassium fluoborate, 0·5 g potassium pyrosulphate, and 0·2 g boric acid, at 510°C for 6 minutes. When the mass is digested with 1:1 hydrochloric acid, diluted, heated, filtered, and washed, free silica remains in the residue [27]. Results are similar to those obtained by heating with phosphoric acid.

C. Other Methods

A different approach to the determination of free silica in silicate dust has been published. It is based on the fact that, from a mixture of methylene blue and basic fuchsin, the silicates adsorb methylene blue preferentially, whereas free silica adsorbs mainly fuchsin. The dust sample is digested in a mixture of hydrochloric and nitric acids, filtered, washed, the residue ignited for 15 minutes at 730–740°C, the dyes added, and the colour of the liquid measured in a photometer against a series of standards [28].

The determination of free silica has been reviewed for silicates [12], rocks and mine dust [29], and for various dusts [30].

2. Elemental Silicon

The determination of elemental silicon, in the presence of silicon carbide, silica, and other siliceous materials has been studied by several workers.

The measurement of the hydrogen evolved when metallic silicon reacts with a solution of sodium hydroxide is the basis of one method. The sample in a pressure bottle may be heated with saturated sodium hydroxide solution at 160–200°C for 40 minutes, cooled rapidly, and the volume of hydrogen measured [31]. Alternatively, the sample may be boiled for 4 hours with 25% sodium hydroxide solution, and the volume of released hydrogen determined [32].

In another procedure, the determination of free silicon is based on the difference between the absorbance of the molybdisilicic acid complex obtained by nitric–hydrofluoric acid attack which dissolves free silica and free and combined silicon, and that of the complex obtained by attack with sulphuric–hydrofluoric acid which dissolves only free silica [33].

It has been stated that silica may be determined in the presence of metallic silicon by reaction at 70°C with 40% fluoboric acid, to yield silicon tetrafluoride which is determined colorimetrically. Silicon is not attacked [34].

Small concentrations of silicon in natural and pollution aerosols can be measured by neutron activation analysis [35].

In complex slags, the silicate phase has been separated by phosphoric acid treatment, free silica by reaction with hydrofluoric acid, elemental silicon by treatment with a mixture of nitric, hydrofluoric and sulphuric acids, and finally silicon carbide by fusion with a mixture of sodium and potassium carbonates [36].

The determination of the forms of silicon in water has been reported [37]. Traces of silicon in the water for high-pressure, steam-generating equipment of modern power stations may be important. "Reactive silicon" includes those forms of silicon, mainly monomeric and dimeric silicic acid, that react with ammonium molybdate in 10 minutes under the conditions of the method. Molybdisilicic acid is reduced by 1-amino–2-naphthol–4-sulphonic acid containing sodium sulphite and potassium metabisulphite.

REFERENCES

1. FURMAN, N. H., ed., *Scott's Standard Methods of Chemical Analysis*, 6th ed., Vol. 1, Princeton, N.J., Van Nostrand, 1962.
2. KAPLAN, E., and FALES, W. T., *Ind. Eng. Chem. Anal. Ed.* **10**, 388 (1938).
3. GURVITS, S. S., and PODGAITS, V. V., *Zavod. Lab.* **14**, 935–8 (1948). *C.A.* **43**, 2547 (1949).
4. SYSOEVA, R. S., *Opredelenie Svobodnoi Dvyokisi Kremniya v Gorn. Porodakh i Rudn. Pyli, Akad. Nauk S.S.S.R., Inst. Gorn. Dela, Sbornik Statei* **1958**, 103–10. *C.A.* **54**, 5334 (1960).
5. LINE, W. R., and ARADINE, P. W., *Ind. Eng. Chem. Anal. Ed.* **9**, 60–3 (1937).
6. JUNG, H., *Naturwissenschaften* **30**, 266–7 (1942). *C.A.* **36**, 5442 (1942).
7. AUBRY, J., and TURPIN, G., *Rev. met.* **47**, 146–7 (1950). *C.A.* **44**, 5266 (1950).
8. SHCHEKATURINA, L. G., and PETRASHEN, V. I., *Opredelenie Svobodnoi Dvuokisi Kremniya v Gorn. Porodakh i Rudn. Pyli, Akad. Nauk S.S.S.R., Inst. Gorn. Dela, Sbornik Statei* **1958**, 54–7. *C.A.* **54**, 5333 (1960).
9. TALVITIE, N. A., *Anal. Chem.* **23**, 623–6 (1951).
10. ZURLO, N., and GRIFFINI, A. M., *Med. lavoro* **45**, 675–91 (1954). *C.A.* **49**, 9440 (1955).
11. COLL, A. G., *Bol. quim. peruano* **3**, No. 17, 5–8 (1955). *C.A.* **50**, 14449 (1956).
12. AVY, A., *Mem. serv. chim. état.* **40**, 345–62 (1955). *C.A.* **52**, 7016 (1958).
13. FAINBERG, S. Y., and BLYAKHMAN, A. A., *Analiz Rud Tsvetnykh Metal. i Produktov ikh Pererabotki* **1956**, No. 12, 111–18. *C.A.* **51**, 6432 (1957).
14. BULYCHEVA, A. I., and MEL'NIKOVA, P. A., *Opredelenie Svobodnoi Dvuokisi Kremniya v Gorn. Porodakh i Rudn. Pyli, Akad. Nauk S.S.S.R., Inst. Gorn. Dela, Sbornik Statei* **1958**, 23–32. *C.A.* **54**, 5332 (1960).
15. RHEDEY, P., and ROBOZ, J., *Banyasz. Lapok.* **11**, 402–6 (1956). *C.A.* **52**, 7935 (1958).
16. DOBROVOL'SKAYA, V. V., *Opredelenie Svobodnoi Dvuokisi Kremniya v Gorn. Porodakh i Rudn. Pyli, Akad. Nauk S.S.S.R., Inst. Gorn. Dela, Sbornik Statei* **1958**, 15–22. *C.A.* **54**, 5332 (1960).
17. TALVITIE, N. A., *J. Am. Hyg. Assoc.* **25**, 169–78 (1964).
18. SHRIVANEK, V., *Sb. Praci Ustavu Vyz. Rud.* **4**, 243–7 (1962). *C.A.* **58**, 6189 (1963).
19. RODIONOVA, A. E., KOBILEV, A. G., and OVCHARENKO, P. P., *Izv. Vysshikh Uchebn. Zavedenii, Khim. i Khim. Tekhnol.* **6**, 518–21 (1963). *C.A.* **60**, 17 (1964).
20. JARZEBOWSKA, J., *Chem. Anal. (Warsaw)* **12**, 835–9 (1967). *C.A.* **68**, 111162 (1968).

21. RABSON, S. R., *J. Chem. Met. Mining Soc. South Africa* **XLV**, 43–57 (1944).
22. FLORENTIN, D., and HEROS, M., *Bull. soc. chim. France* **1947**, 213–15. *C.A.* **41**, 7308 (1947).
23. LITEANU, C., STRUSIEVICI, C., and RUSU, G., *Acad. Rep. Populare Romine, Studii Cercetari Chem.* **3**, 61–6 (1955). *C.A.* **50**, 9228 (1956).
24. ASTAF'EV, V. P., *Opredelenie Svobodnoi Dvuokisi Kremniya v Gorn. Porodakh i Rudn. Pyli, Akad. Nauk S.S.S.R., Inst. Gorn. Dela, Sbornik Statei* **1958**, 51–3. *C.A.* **54**, 5333 (1960).
25. POLEZHAEV, N. G., *Opredelenie Svobodnoi Dvuokisi Kremniya v Gorn. Porodakh i Rudn. Pyli, Akad. Nauk S.S.S.R., Inst. Gorn. Dela, Sbornik Statei* **1958**, 33–43. *C.A.* **54**, 5333 (1960).
26. TROTSENKO, D. D., *Lekarstv. Syr'evye Reswisy Irkutskoi Oblasti, Irkutsk, Sb.* **1961**, 200–5. *C.A.* **60**, 11379 (1964).
27. JAYPRAKASH, K. C., and MAJUMDAR, S. K., *Indian J. Technol.* **4**, 378 (1966). *C.A.* **66**, 61574 (1967).
28. BERKOVICH, M. T., *Opredelenie Svobodnoi Dvuokisi Kremniya v Gorn. Porodakh i Rudn. Pyli, Akad. Nauk S.S.S.R., Inst. Gorn. Dela, Sbornik Statei* **1958**, 69–71. *C.A.* **54**, 5333 (1960).
29. SKOCHINSKII, A. A. ed., *Opredelenie Svobodnoi Dvuokisi Kremniya v Gorn. Porodakh i Rudn. Pyli, Sbornik Statei i Instruktsii. Izdatel'. Akad. Nauk S.S.S.R., Moscow,* 1958. *C.A.* **53**, 19707 (1959).
30. DREES, W., *Silikat Tech.* **14**, 229–35 (1963). *C.A.* **60**, 15107 (1964).
31. SHINKAI, S., *Japan Analyst* **2**, 433–5 (1953). *C.A.* **48**, 6315 (1954).
32. BASTIN-MERKEMAN, M. J., and GOFFAUX, R., *Bull. soc. chim. France* **1963**, 93–9. *C.A.* **58**, 11953 (1963).
33. SERRINI, G., and LEYENDECKER, W., *Met. Ital.* **64**, 129–37 (1972). *C.A.* **77**, 109038 (1972).
34. HOUDA, M., and JOST, F., *Collection Czech. Chem. Commun.* **31**, 776–81 (1966). *C.A.* **64**, 11859 (1966).
35. VAN GRIEKEN, R., and DAMS, R., *Anal. Chim. Acta* **63**, 369–81 (1973).
36. CHALUS, F., *Hutnicke listy* **16**, 587–8 (1961). *C.A.* **56**, 929 (1962).
37. WILSON, A. L., *Analyst* **90**, 270–7 (1965).

Sulphur

In rocks, minerals, ores, and concentrates, sulphur may occur in the form of sulphide or sulphate. Flue gases may contain sulphur as dioxide or trioxide; in other effluent gases hydrogen sulphide may be present. A determination of the phases of sulphur is important in mining and metallurgy, power generation, and many industrial operations.

1. Sulphide and Sulphate Sulphur in Ores, etc.

Many materials contain sulphides and sulphates, both of which may include water-soluble forms like sodium sulphide or cobalt sulphate, and water-insoluble compounds like copper sulphide or lead sulphate. When soluble sulphides and sulphates occur together, the former may be expelled as hydrogen sulphide by acidifying with hydrochloric acid and boiling. The usual analysis for sulphur on the residual solution represents sulphate sulphur. Total sulphur is determined on a separate sample by an oxidizing treatment with bromine to convert sulphides to sulphates; total sulphur − sulphate sulphur = sulphide sulphur.

If an insoluble sulphate must be differentiated from an insoluble sulphide, there are several alternatives for the analyst. In one procedure, sulphur is evolved as hydrogen sulphide from a sulphide by the addition of hydrochloric acid, in the presence of either soluble or insoluble sulphate, absorbed in ammoniacal cadmium chloride solution and finally determined iodometrically. A representative example of this important technique is outlined below.

Weigh 0·5–5 g of −100 mesh sample into a 500-ml Erlenmeyer flask which is fitted with a dropping funnel and a reflux condenser; the top of the latter is connected to a litre flask which serves as the absorption vessel. Add 100 ml hydrochloric acid rapidly through the dropping funnel, heat the Erlenmeyer flask, and boil until the sample is decomposed. Solution of pure nickel may be accelerated by addition of a few drops of platinic chloride solution. Remove the absorption flask, to which had been originally added 50–100 ml ammoniacal cadmium chloride solution diluted with water to about 400 ml, and add 10–25 ml of standard potassium iodate/potassium iodide solution. Acidify the solution with about 50 ml hydrochloric acid, stopper, and shake the flask vigorously. Add starch solution, and titrate the excess iodine with standard sodium thiosulphate.

In the presence of selenium, a part of which is evolved as hydrogen selenide, insert between the reflux condenser and the absorption flask a bottle containing 200 ml of an acid zinc–potassium chloride solution; this retains selenium but allows hydrogen sulphide to pass through.

Ammoniacal cadmium chloride solution. Dissolve 22·5 g cadmium chloride in 1 litre of 1:1 ammonium hydroxide; 50 ml of this solution will absorb the hydrogen sulphide from a sample containing 0·17 g sulphur.

Potassium iodate/potassium iodide solution. Dissolve 5·5625 g potassium iodate and 24 g of potassium iodide in water and dilute to 1 litre; 1 ml = 0·0025 g sulphur. Standardize this solution against arsenious acid; 25 ml of 0·1N arsenious acid is equivalent to 16·03 ml of the potassium iodate/potassium iodide solution. A 0·1N arsenious acid solution is made by dissolving 4·946 g pure arsenic trioxide in 15 g sodium carbonate dissolved in a minimum quantity of water, and diluting to 1 litre.

Sodium thiosulphate solution. Dissolve 38·71 g $Na_2S_2O_3.5H_2O$ in 1 litre; 1 ml = 1 ml of the above potassium iodate/potassium iodide solution.

Acid zinc–potassium chloride solution. Mix 10 g of zinc with water, and dissolve in a slight excess of hydrochloric acid. Add 50 g potassium chloride and dilute to 1 litre. Add a solution of potassium hydroxide until a yellow colour is obtained with thymol blue, then add dilute hydrochloric acid until the colour is red [1].

In an alternative procedure for ores and similar materials, boil 1–3 g of —100 mesh sample for 15 minutes with 50 ml of 10% sodium carbonate solution, filter and wash thoroughly with hot water. The filtrate contains the sulphur derived from sparingly soluble or insoluble sulphates like calcium or lead, and the precipitate contains the insoluble metal sulphides. Oxidation of copper sulphides and most other base metal sulphides is negligible [2]. Pyrrhotite, however, becomes more reactive in an alkaline solution, resulting in a slow dissolution of sulphur; if a substantial quantity of this mineral is present, some trials on a minimum boiling time, or a correction for dissolved sulphur, may be required.

A separation of the sulphate sulphur of anhydrite or gypsum from barite and from the sulphide sulphur of ores has been published [3]. To 1–10 g of —200 mesh sample add 500 ml of 10% ammonium chloride solution having a neutral reaction. Allow to stand at room temperature for 4 hours, agitating briskly for a few seconds every 15 minutes. Filter through Whatman No. 42 paper, wash thoroughly with cold water, and determine sulphur on the filtrate by the conventional procedure. This represents the sulphur from anhydrite or gypsum; barite and the sulphides of cobalt, copper, iron, lead, and zinc remain virtually insoluble. The comparatively rare strontium sulphate mineral, celestite, is partially soluble in this solution.

The differentiation of pyrite from pyrrhotite by treatment with a hot solution of 2 parts water and 1 part hydrochloric acid has already been discussed in the chapter on iron. The same technique can be used to distinguish these forms of sulphur.

In another method of distinguishing pyrite sulphur from pyrrhotite sulphur, in the presence of barite, the detailed directions take the following form. For pyrite sulphur, moisten a 1–5 g sample in a platinum dish with a small amount of water, add 30 ml of 1:2 hydrochloric acid, and evaporate at 50–60°C. Add 40 ml of 1:1 hydrochloric acid and 1–2 g of hydroxylamine, warm 10–15 minutes, filter, and wash with 0·1% hydrochloric acid until all iron is removed. Dissolve the precipitate on the filter paper in 30 ml of nitric acid, add 50 ml of hot water to the solution, and filter. Wash with 0·1% hydrochloric acid until the iron is removed. Pass the filtrate through a column of alumina at 5 ml/minute, and wash with 70 ml of 1:20 hydrochloric acid and then with 50 ml of water. Elute the adsorbed sulphate with 5 ml of 1:10 ammonium hydroxide, and wash the column with 50 ml of 1:100 ammonium hydroxide, collecting both solutions. Acidify the total solution with hydrochloric acid to give a concentration of 1 ml per 100 ml of solution, and determine the sulphur by the conventional barium sulphate gravimetric method. To convert the sulphur content to pyrite, use the factor 1·871.

To determine barite sulphur, to a 0·5–1 g sample add 50 ml of 1:2 hydrochloric acid and 2 g sodium fluoride; warm until all black particles disappear. Keep the volume of solution constant. Filter, wash with 0·1% hydrochloric acid, and then with water until all chlorine is removed. Ignite in a platinum crucible, weigh, treat with hydrofluoric acid, ignite, and weigh again [4].

For pyrrhotite sulphur, moisten a 0·1–3 g sample in the flask of a Schulte apparatus with water, and add a freshly prepared solution of stannous chloride containing 4 g stannous chloride in 100 ml of 1:1 hydrochloric acid. Warm until evolution of hydrogen sulphide ceases, absorbing the latter in 2·5% cadmium acetate solution. When the reaction stops, add to the absorption flask containing cadmium sulphide 10 ml of standard iodine solution and 5–7 ml of 2:3 hydrochloric acid. Titrate with standard sodium thiosulphate solution in the presence of starch solution [4].

Sulphate and sulphide sulphur in rocks and minerals have been differentiated by another technique. Total sulphur was determined as usual, following an initial oxidizing treatment to convert all sulphur to the sulphate form. On a separate sample, digestion with acidified barium chloride in an inert atmosphere, in the presence of calcium chloride to precipitate any sulphide that may be liberated, will convert sulphate to barium sulphate. Sulphide sulphur is found by difference [5].

Another procedure for rocks is based on the fact that sulphide sulphur is evolved as hydrogen sulphide when the sample is heated with phosphoric acid, and can be determined by absorption and iodometric titration. If the sample is heated with a mixture of stannous chloride and phosphoric acid, both sulphide and sulphate sulphur are evolved as hydrogen sulphide, and determined in the usual manner [6].

A chemical phase analysis of the different forms of sulphur in the system lead–sulphur–oxygen has been published [7]. To a 0·25–0·5 g sample add 20–25 ml of 5% sodium hydroxide, boil 3–5 minutes if free sulphur is absent; if sulphur is present stir with 10% sodium hydroxide at room temperature for 30 minutes. Dilute to 100 ml, filter, wash with small quantities of 1% sodium hydroxide, and make up to 200 ml. To the precipitate add about 30 ml water and determine sulphide sulphur by decomposition with hydrochloric acid, distilling the hydrogen sulphide into standard iodine solution and back-titrating the excess iodine with standard sodium thiosulphate.

Determine the total sulphur after the alkaline treatment of a second sample. Take three aliquots from the alkaline filtrate. To the first, add an excess of 0·05N iodine solution, let stand a few minutes, acidify with hydrochloric acid, and titrate with sodium thiosulphate to obtain $SO_3^{--} + S_2O_3^{--}$. To the second aliquot, add 4–5 ml of 40% formaldehyde, and neutralize dropwise with acetic acid; add an excess of iodine solution and titrate with sodium thiosulphate to obtain $S_2O_3^{--}$. To the third aliquot if $S_2O_3^{--}$ is absent, boil with hydrochloric acid to decompose SO_3^{--}, and determine the remaining sulphate as barium sulphate. If $S_2O_3^{--}$ is present, the third aliquot is subjected to a more lengthy determination of total sulphur. Consequently, $S_2O_3^{--}$ is determined from the second aliquot by $SO_3^{--} = (S_2O_3^{--} + SO_3^{--}) - S_2O_3^{--}$, where $(S_2O_3^{--} + SO_3^{--})$ is determined from the first aliquot. Sulphate is obtained by direct analysis of the third aliquot, or by calculation. Sulphur = $S_{\text{total}} - (S_2O_3^{--} + SO_3^{--})$ where S_{total} is determined by analysis of the residue of the first sample after alkaline treatment, and S is calculated by S = S_{total} — sulphide where S_{total} is determined in the residue from the alkaline attack of the second sample [7].

A determination of the forms of sulphur in an oil shale has been published [8]. Sulphate sulphur is extracted with 10% perchloric acid and determined by the conventional gravimetric method. Pyrite sulphur is reduced to sulphide by reaction with lithium aluminium hydride, and the hydrogen sulphide evolved by acidification is determined by titration of the free acid formed in a neutral cadmium sulphate solution. Organic sulphur is measured gravimetrically by an Eschka fusion of the residue after the mineral sulphur forms are extracted.

The forms of sulphur in the low content of this element in electrolytic cobalt metal can be differentiated by standard techniques, using a large sample. Total sulphur is found by burning a 10-g sample in a stream of air at 1460°C in a tubular furnace, absorbing the sulphur dioxide in acidified starch iodide solution, and continuously titrating with 0·001M potassium iodate solution. Sulphide sulphur is measured by boiling a large sample in hydrochloric acid, sweeping out the hydrogen sulphide in a stream of nitrogen, passing the gases through an acid trap, absorbing in ammoniacal zinc chloride, and finishing by the usual iodine–thiosulphate titration [9].

2. Sulphur in Gases

The importance of the oxides of sulphur in flue gases has resulted in numerous investigations on their determination. Sulphur trioxide can be separated from 10–100 times as much sulphur dioxide by passing the gas through an 80% solution of isopropyl alcohol in water [10].

A common procedure is carried out as follows [11]. The gas is absorbed in standard sodium hydroxide solution containing benzyl alcohol, sodium sulphate, benzaldehyde, and paraaminophenol hydrochloric acid. It is then titrated with standard hydrochloric acid to an endpoint of *p*H 4·1 by means of a *p*H meter. The sulphur trioxide is then determined by adding benzidine solution and hydrochloric acid, filtering, boiling the precipitate and paper with water, titrating past the phenolphthalein endpoint with standard sodium hydroxide, cooling, and titrating to the endpoint with standard hydrochloric acid. The sulphur dioxide is calculated from the first hydrochloric acid titration and the titration of benzidine sulphate.

Sulphur in the form of sulphuric acid mist is removed by a sintered glass filter, before measuring sulphur dioxide by oxidation with hydrogen peroxide, and titration with sodium hydroxide to the methyl red endpoint [12].

Another procedure for oxides of sulphur recommends two absorbers, both containing 250 ml of 0·25N hydrogen peroxide solution previously standardized against potassium permanganate. Separate aliquots are titrated with standard potassium permanganate and with standard sodium hydroxide solution. The permanganate titration gives the content of sulphur dioxide, and the sodium hydroxide titration measures sulphur dioxide + sulphur trioxide [13].

Sulphur oxides in stack gases have been measured by the following procedure. Sulphur trioxide is selectively absorbed in 80% isopropyl alcohol by three absorbers in an ice slurry. To an aliquot is added a few drops of thorin indicator, and the solution is titrated with standard barium chloride solution to the pink endpoint. For total sulphur oxides, the gas is passed through two absorbers of 3% hydrogen peroxide and 0·2N sodium hydroxide. The solution is boiled to remove peroxide, acidified with hydrochloric acid, boiled, cooled, and diluted to volume. An aliquot is passed through a Dowex 50 resin column, isopropyl alcohol and thorin added, and titration is carried out with standard barium chloride solution to the pink endpoint [14].

Sulphur trioxide in flue gas has been absorbed in 80% isopropyl alcohol, and titrated with standard barium perchlorate in 80% isopropyl alcohol to the appearance of a pink colour with 0·04% thorin indicator. Sulphur dioxide at a concentration of 60 μg/ml in the isopropyl alcohol solution does not interfere [15].

Another proposal for sulphur oxides in flue gases is to draw the gas at 250°C into a condenser at 60–90°C, wash the sulphuric acid from the collector with 5% aqueous isopropyl alcohol of *p*H 4·6, and titrate with standard sodium hydroxide, using bromophenol blue as indicator. Pass the residual gas, freed of sulphur trioxide, into 3% hydrogen peroxide at *p*H 4·6, and titrate the solution with standard sodium hydroxide, using bromophenol blue indicator [16].

If a gas contains hydrogen sulphide and sulphur dioxide, the components can be separated by passing the gas through an absorber containing a neutral hydrogen peroxide solution to oxidize sulphur dioxide, and then through another absorber containing alkaline hydrogen peroxide, where hydrogen sulphide is oxidized. The sulphuric acid formed in each absorber is then titrated with standard sodium hydroxide [17].

REFERENCES

1. Young, R. S., *Chemical Analysis in Extractive Metallurgy*, London, Charles Griffin, 1971.
2. Young, R. S., *Industrial Inorganic Analysis*, London, Chapman and Hall, 1953.
3. Young, R. S., and Hall, A. J., *J. Soc. Chem. Ind.* **66**, 375 (1947).
4. Puzankova, N. V., and Dymov, A. M., *Izv. Vyssh. Ucheb. Zaved. Chern. Met.* **11**, 176–80 (1968). *C.A.* **69**, 32751 (1968).
5. Vlisidis, A. C., *U.S. Geol. Surv. Bull.* No. 1214–D, 1966.
6. Nagashima, S., Yoshida, M., and Uzawa, T., *Bull. Chem. Soc. Jap.* **45**, 3446–51 (1972). *C.A.* **78**, 23533 (1973).
7. Grishankina, N. S., and Margulis, E. V., *Sb. Nauchn. Tr. Vses. Nauchn.-Issled. Gorno-Met. Inst. Tsvetn. Metal* No. **9**, 167–73 (1965). *C.A.* **63**, 17141 (1965).

8. SMITH, J. B., YOUNG, N. B., and LAWLOR, D. L., *Anal. Chem.* **36**, 618–22 (1964).
9. SMITH, G. A., and MACLEOD, D. S., *Trans. Institution Mining Metallurgy* **79**, C41–53 (1970).
10. CORBETT, P. F., *J. Soc. Chem. Ind.* **67**, 227–30 (1948).
11. BERK, A. A., and BURDICK, L. R., *U.S. Bur. Mines Rept. Invest.* 4618, 1950.
12. LOMBARDO, J. B., *Anal. Chem.* **25**, 154–60 (1953).
13. PANNETIER, G., and MELTZHEIM, C., *Bull. Soc. Chim. France* **1955**, 1186–7. *C.A.* **50**, 1525 (1956).
14. SEIDMAN, E. B., *Anal. Chem.* **30**, 1680–2 (1958).
15. FIELDER, R. S., and MORGAN, C. H., *Anal. Chim. Acta* **23**, 538–40 (1960).
16. CHORY, J. P., *Brennstoff-Waerme Kraft* **14**, 601–3 (1962). *C.A.* **58**, 10728 (1963).
17. VAN STRATEN, H. A. C., *Anal. Chim. Acta* **14**, 325–8 (1956).

Tellurium

A method has been published for the determination of oxidized tellurium in powdered metallic tellurium.

When the sample is allowed to react at room temperature for 15–20 minutes with 40% sulphuric acid, tellurium dioxide dissolves whereas metallic tellurium is unattacked [1].

If the sample is in the liquid state, which happens frequently in leach plants or electrolytic refineries, tellurium (IV) can be separated from tellurium (VI) by thin-layer chromatography on cellulose powder with acetone/3N hydrochloric acid/2% tartaric acid [2].

REFERENCES

1. RYBAKOV, B. N., MASLOVA, G. V., and SINYAGOVSKAYA, L. A., *Zavod. Lab.* **35**, 268–9 (1969). *C.A.* **71**, 9403 (1969).
2. MEKRA, H. P., MITTAL, I. P., and JOHRI, K. N., *Chromatographia* **4**, 532–4 (1971). *C.A.* **76**, 94122 (1972).

Tin

A number of publications have appeared on the phase analysis of natural and processed forms of tin.

Mixtures of tin and its oxides have been investigated. In one method, the sample is treated with a solution of bromine in chloroform to dissolve metallic tin, leaving stannous and stannic oxides intact [1].

In another procedure, the sample is shaken for 1–4 hours in an atmosphere of carbon dioxide with a neutral 5% ferric sulphate solution, filtered, and metallic tin is calculated from the ferrous iron content of the filtrate, which is determined by standard potassium permanganate. The residue of tin oxides is heated for 1 hour at 100°C with a solution of 4% oxalic acid and 4% ammonium oxalate, and filtered. Stannous oxide dissolves, leaving stannic oxide in the final residue [1, 2].

Other workers have used standard potassium dichromate to determine ferrous iron produced from a ferric solution by metallic tin, in the presence of tin dioxide [3].

Studies have been made on the forms of tin in ores. Tin sulphide can be determined by stirring 0·5 g of —100 mesh sample for 2·5 hours at room temperature in a solution containing 25 ml carbon tetrachloride, 10 ml bromine, and 1 g of sulphur. After filtration and washing, tin sulphide is in the filtrate; stannous oxide and primary cassiterite are in the residue [4].

In the extractive metallurgy of tin, a chloridizing roast may yield a mixture of cassiterite, stannous chloride, stannous oxide, tin silicate, and metallic tin. It has been recommended to extract, successively, metallic tin with a tartaric acid solution of cupric chloride, and stannous oxide and silicate by the action of hot hydrochloric acid. Cassiterite remains in the final residue [5].

The mineral stannite and various oxidized colloidal forms of tin are dissolved by boiling first with hydrochloric acid and potassium chlorate, and then with nitric acid; cassiterite is unattacked and is left in the residue [6].

A phase analysis of tin ores containing hydrated tin compounds has been reported. When a finely ground sample is treated with a solution of bromine in ethanol/carbon tetrachloride, stannite and cassiterite are dissolved, leaving tin hydrates in the residue. In another sample, treatment with hot sulphuric acid leaves cassiterite unattacked. Determination of tin in all filtrates and residues gives a close indication of the ore composition [7].

A different technique has been employed in a phase analysis of tin slags. When a stream of chlorine is passed over the sample in a tube heated to 100–110°C, metallic tin and sulphide tin are removed as volatilized chlorides whereas oxidized tin remains in the residue [8].

A determination of total, free, and alloyed tin in tin plate has been published. Total tin was found by the conventional iodimetric procedure. Free tin was determined by digestion in a solution containing 80 g lead acetate and 135 g sodium hydroxide in 1 litre, followed by filtration, and analysis of the filtrate for tin. Alloyed tin was unattacked and remained in the residue, but was usually calculated by difference between total and free tin [9].

A novel differentiation of metallic from oxide tin is described in a method for oxygen content of tin powder. The sample is fused with colophony at 260–275°C. Pure tin forms a regulus below the slag of oxygen-containing compounds [10].

If the sample is received in liquid form, tin (II) can be separated from tin (IV) by thin-layer chromatography, using a butyl alcohol/acetic acid mixture [11].

A phase analysis of tin compounds in zinc concentrates and their re-processing products has been described [12]. To 0·5–1 g sample add 50 ml of 0·5% cupric chloride solution containing 1·7% tartaric acid, stir for 30 minutes, and filter. Wash the residue (I) with tartaric acid and then with water, and retain. To the filtrate add 10 ml nitric acid, 15 ml 1:1 sulphuric acid, and evaporate to fumes. Cool, dilute, heat, add 10 ml of 0·5% ferric chloride solution, precipitate ferric and stannous hydroxides with ammonium hydroxide, and filter. Dissolve the residue in hydrochloric acid, and repeat the precipitation. Place the precipitate in an iron crucible, dry, ignite, and fuse with 4–5 g of sodium peroxide. Leach with hot water, add 20 ml of hydrochloric acid, filter, and wash with dilute hydrochloric acid. Treat the solution with 0·5 g of mercuric chloride and 1 g sodium thiosulphate, boil for 3 minutes, and cool. Add 1–2 g of potassium iodide, and titrate tin with a 0·02N solution of iodine in the presence of starch solution.

For the determination of tin as the sum of tin silicate and stannous oxide, proceed as follows [12]. To the residue (I) above, add 50 ml of 0·7% potassium fluoride in 0·5% sulphuric acid. Mix for 30 minutes, filter, and wash 5–6 times with the same reagent. Keep this residue (II) for the following analysis. To the filtrate add 10 ml of 0·5% ferric chloride solution and continue as above. For the determination of tin as sulphide, treat residue (II) with 40 ml hydrochloric acid, cover the flask with a funnel, and boil for 30 minutes. Add 40 ml of hot water, decant, wash 2–3 times with hot 1:1 hydrochloric acid, add 30 ml of hydrochloric acid, dilute with water, and filter. Repeat the operation and keep the residue for further analysis. To the solution add 10 ml of nitric acid, 15 ml of 1:1 sulphuric acid and analyse as above. To determine cassiterite, dry the residue retained above, burn to ash, and determine as described.

REFERENCES

1. Vasil'ev, V. V., and Novikov, R. N., *Uchenye Zapiski Leningrad. Gosudarst. Univ.* No. **163**, *Ser. Khim. Nauk.* No. **12**, 15–27 (1953). *C.A.* **49**, 2940 (1955).
2. Gauzzi, F., *Ann. chim.* **47**, 1316–20 (1957). *C.A.* **52**, 7017 (1958).
3. Spauszus, S., and Lanz, G., *Chem. Tech.* **14**, 111–13 (1962). *C.A.* **57**, 1541 (1962).
4. Zverev, L. V., and Petrova, N. V., *Zavod. Lab.* **23**, 1403–5 (1958). *C.A.* **54**, 10652 (1960).
5. Filippova, N. A., and Korosteleva, V. A., *Analiz Rud Tsvetnykh Metal. i Produktov ikh Pererabotki, Sbornik Nauch. Trudov* No. **14**, 143–54 (1958). *C.A.* **53**, 13873 (1959).
6. Vlasova, G. M., *Uch. Zap., Tsentr. Nauchn.-Issled. Inst. Olovyan. Prom.* **1**, 14–17 (1965). *C.A.* **66**, 8118 (1967).
7. Karapetyan, E. T., and Sharko, E. D., *Obogashch. Rud* **15**, 102–4 (1970). *C.A.* **73**, 51925 (1970).
8. Filin, N. A., and Mamontov, N. F., *Trudy Vsesoyuz. Nauch. Inzh.-Tekh. Obshchestva Metallurgov* **2**, 199–203 (1954). *C.A.* **52**, 18074 (1958).
9. Maguid, A., *Rev. Fac. Ing. Quim., Univ. Nac. Litoral* **36**, 281–303 (1967). *C.A.* **71**, 98041 (1969).
10. Gumilevskaya, G. P., Chekanova, V. D., Guseva, K. S., Tashkova, Z. I., and Murilina, A. I., *Tr. Vses. Nauchn.-Issled. Proektno-Tekhnol. Inst. Elektrougol'n Izdelii* **1970**, 119–23. *C.A.* **75**, 136745 (1971).
11. Mekra, H. P., Mittal, I. P., and Johri, K. N., *Chromatographia* **4**, 532–4 (1971). *C.A.* **76**, 94122 (1972).
12. Filippova, N. A., *Analiz Rud Tsvetnykh Metal. i Produktov ikh Pererabotki* **1956**, No. 12, 70–8. *C.A.* **51**, 14472 (1957).

Titanium

A few publications have appeared on the phase analysis of titanium in ores, metals, slags, and alloys.

Ilmenite in the presence of titaniferous magnetite is determined by treatment with 8% hydrochloric acid, which dissolves magnetite but is without effect on ilmenite. Sphene and perovskite can be separated from titaniferous magnetite by the solubility of the latter in phosphoric acid [1].

In another study, the recommendations are as follows. Digest 0·2 g of −100 mesh ore with 20 ml of 8N hydrochloric acid containing 0·4 g of sodium fluoride, filter, and wash. The filtrate contains titanium from sphene, ilmenite, and titanomagnetite; the residue contains titanium from rutile and leuxene. To determine titanium in sphene and ilmenite, reduce 0·2–0·4 g in hydrogen at 800°C for 4 hours. Cool, treat with 25 ml of N hydrochloric acid at 60°C to dissolve iron. After 30 minutes add 50 ml hydrochloric acid and 0·4–1 g of sodium fluoride. Heat for 1 hour, dilute, filter and wash. In the filtrate is the titanium from sphene; the titanium in the residue is derived from ilmenite [2].

A phase analysis of titanium metal has been published. Total titanium was found by conventional chemical procedures. Metallic titanium was calculated from the weight increase after ignition at 1200°C for 30 minutes. Titanium carbide was measured by the usual combustion technique, and titanium nitride was obtained by the familiar Kjeldahl method. Titanium oxide was found by difference [3].

The reduced titanium, or in other words the sum of the lower oxides, in slags has been determined. Boil the sample in 10% sulphuric acid for 15–20 minutes in an atmosphere of carbon dioxide. Add a measured amount of vanadium pentoxide to oxidize the lower titanium oxides, 30 ml of 70% sulphuric acid, cool, dilute, and titrate with standard ferrous ammonium sulphate solution, using barium diphenylamine sulphonate as indicator [4].

Titanium and its dioxide have been differentiated in an alloy steel by dissolving the sample in 33% nitric acid, filtering, and washing. Titanium dioxide remains in the residue, whereas the remainder of the titanium is found in the filtrate [5].

A method for distinguishing titanium oxide, carbide, and nitride in nickel–titanium alloys has been reported. Treat 1 g of the finely pulverized sample with 200 ml of methyl acetate containing 20 ml of bromine at a temperature below 20°C for 1 hour, filter, and wash with methanol. Treat the residue with a few drops of hydrofluoric acid, dilute, filter, and wash. This residue contains titanium oxide. Treat the residue, which remains undissolved after hydrofluoric acid addition, with 25 ml of 1:1 nitric acid and 5 ml of hydrofluoric acid for 30 minutes on a water bath. Dilute, warm, filter, wash, and determine in the filtrate the titanium derived from carbide and nitride. In a separate sample of 5 g dissolve the metals by digesting below 20°C for 1 hour with a mixture of 20 g of bromine in 200 ml of methanol. Filter, wash with methanol, and treat the residue with 7 ml of hydrofluoric acid and 30% hydrogen peroxide on a boiling water bath for 1 hour. Determine the ammonia formed in the solution by the usual Kjeldahl method, and calculate the amount of titanium present as nitride. Titanium carbide is found by difference.

REFERENCES

1. Perlov, P. M., and Katsnel'son, E. M., *Obogashchenie Rud* **4**, No. 6, 11–13 (1959). *C.A.* **55**, 26844 (1961).
2. Fedorova, M. N., and Krivodubskaya, K. S., *Zavod. Lab.* **30**, 515–18 (1964). *C.A.* **61**, 4943 (1964).
3. Goto, H., and Takeyama, S., *Japan Analyst* **1**, 7–9 (1952). *C.A.* **47**, 1002 (1953).
4. Chernomordik, E. M., *Zavod. Lab.* **11**, 796–803 (1945). *C.A.* **40**, 7058 (1946).
5. Tsinberg, S. L., *Zavod. Lab.* **6**, 358 (1937). *C.A.* **32**, 4100 (1938).
6. Tsukahara, I., and Yabuki, E., *Nippon Kinzoku Gakkaishi* **36**, 66–72 (1972). *C.A.* **76**, 54060 (1972).

Tungsten

A phase analysis of tungsten ores and concentrates has been published [1]. For a mixture of tungstite, scheelite, wolframite, and hubnerite, successive treatments with ammonium hydroxide and oxalic acid extracted tungstite first, and then scheelite; wolframite and hubnerite remained in the residue. Materials rich in tungstite were given four treatments with ammonium hydroxide, and the remaining ammoniacal solutions were analysed separately. Materials rich in scheelite were treated twice with oxalic acid solution. The tungsten of all four minerals could be determined satisfactorily, but for materials containing both wolframite and hubnerite the accuracy of the determination for each of these minerals was lower.

REFERENCE

1. Solntsev, N. I., and Leont'eva, K. D., *Analiz Rud Tsvetnykh Metal. i Produktov ikh Pererabotki, Sbornik Nauch. Trudov* **1958**, No. 14, 155–68. *C.A.* **53**, 13872 (1959).

Uranium

Several phase analyses of uranium compounds have been reported.

The oxidation state of uranium in apatite and phosphorite deposits has been investigated [1]. Carbonate–fluorapatite is dissolved in 1·2M hydrochloric acid containing hydroxylamine hydrochloride. The latter suppresses oxidation of uranium (IV) without reducing uranium (VI). Uranium (IV) is precipitated with cupferron, using titanium as a carrier. The ignited precipitate is dissolved in nitric acid; the uranium is extracted with ethyl acetate and determined fluorimetrically. Formation of a uranium (IV)–fluorine complex probably causes about 20% low recovery of uranium (IV) from fluorapatites dissolved in hydrochloric acid and hydroxylamine hydrochloride. Nearly all sedimentary and igneous apatites were found to contain some uranium (IV). The uranium (IV) content of sedimentary apatites ranges from 3 to 91% of the total uranium; the uranium (IV) content in igneous apatites represents 10–66% of their total uranium.

Metallic uranium has been determined in the presence of uranium oxide by the selective dissolution of the former in ethyl acetate containing 4M bromine. The acids produced in side reactions of bromine with the solvent are neutralized by magnesium oxide. As far as possible, the sample should be dry and the reagents anhydrous; water causes some solution of uranium oxide [2].

In another publication, the differentiation of uranium metal from uranium oxides and carbides has been described. Allow 1 g of sample to react for 30 minutes with 100 ml of ethyl acetate containing 20% bromine, filter, and wash. Metallic uranium is in the filtrate; oxides and carbides remain in the residue [3].

A rapid determination of uranium dioxide in uranium phosphides has been recommended. The phosphide compounds UP and U_3P_4 are dissolved in a solution of ethyl acetate and hydrochloric acid, and the remaining UO_2 is oxidized in air to U_3O_8 [4].

REFERENCES

1. Clarke, Jr., R. S., and Altschuler, Z. S., *Geochim. et Cosmochim. Acta* **13**, 127–42 (1958).
2. Brunzie, G. F., Johnson, T. R., and Steunenberg, R. K., *Anal. Chem.* **33**, 1005–6 (1961).
3. Ashbrook, A. W., *Analyst* **87**, 751–4 (1962).
4. Driscoll, J. L., and Evans, P. E., *Analyst* **93**, 403–5 (1968).

Zinc

A great deal of work has been done to distinguish by chemical analysis the forms of zinc in natural and processed materials.

1. Zinc Oxide

A. In Ores, Flue Dusts, etc.

In the extractive metallurgy of zinc, oxidized ores require a treatment differing from that for sulphides, and it is important to differentiate these forms. Various procedures have been proposed for this purpose.

With some simple ores, boil 1 g of –100 mesh sample for 10 minutes with 25 ml of 25% ammonium chloride solution and 10 ml of saturated ammonium acetate. Filter, and wash thoroughly with hot water; oxide zinc passes into the filtrate whereas zinc sulphide remains in the residue [1].

Several variations of the solvent have been proposed. In one, treatment is for 1 hour at 60°C with a solution containing 32 g ammonium chloride, 80 ml of ammonium hydroxide, and 100 ml of water [2].

In another, free zinc oxide is extracted from ferrite by treatment with 12% ammonium acetate or 10% sodium hydroxide for 1 hour at 90°C [3].

One publication recommends heating a mixture of zinc oxide, zinc silicate, and quartz for 1 hour with a solution containing 32 g ammonium chloride in 200 ml of 10% ammonium hydroxide; after filtration zinc oxide is in the filtrate [4].

Other workers have suggested shaking a 2-g sample at 15-minute intervals for 2 hours, or continuously for 10 minutes, with 100 ml of 30% ammonium acetate to dissolve zinc oxide in zinc residues [5] and dusts [6].

One recommendation favours ammoniacal ammonium carbonate solution to dissolve zinc oxide in the presence of ferrite and sulphide [7].

The most successful selective solvent for complex non-sulphide zinc ores in the author's experience is 2% by volume sulphuric acid saturated with sulphur dioxide. The oxidized zinc minerals willemite, hemimorphite, hopeite, parahopeite, tarbuttite, zincite, descloizite, and smithsonite are practically completely soluble, whereas zinc sulphide is nearly insoluble [8]. The simple procedure is outlined below.

Weigh 0·5–1 g of –200 mesh sample into a 300 ml Erlenmeyer flask fitted with a Bunsen valve. Add 50 ml of 2% by volume sulphuric acid saturated with sulphur dioxide. Stopper the flask and allow to stand in a warm place at 30–40°C for 1 hour, swirling the flask for a few seconds every 10 minutes. Filter and wash thoroughly with hot water. The filtrate contains the zinc from oxidized minerals; zinc sulphide remains in the residue.

The method is equally satisfactory for mixed lead–zinc minerals, ores, and concentrator products in the presence of iron, vanadium, manganese, copper, etc. With these conditions of particle size and agitation, there is apparently no tendency for zinc minerals to be protected by a layer of insoluble lead sulphate. For very refractory silicate minerals, it may be necessary to add a few drops of hydrofluoric acid to the leach solution. This has been found to be without solvent action on zinc sulphide under these conditions [8].

B. In Powder, Dust, etc.

Finely divided zinc, called zinc powder or zinc dust, is a familiar article of commerce; zinc oxide likewise has many uses. In both these substances it is important to know the content of zinc oxide as well as total zinc.

In general, the determination of zinc oxide in zinc powder or dust, or in commercial zinc oxide, can be carried out by the solution of zinc oxide in solvents such as ammonium acetate, ammoniacal ammonium chloride, and others used for ores, flue dusts, and metallurgical residues. As noted earlier, zinc sulphide and metallic zinc are not attacked and remain in the residue. The procedure below is typical.

To a 2-g sample add 100 ml of 30% ammonium acetate, stir for 10 minutes, and allow to stand for 1 hour. Zinc oxide is dissolved, leaving metallic zinc in the residue. If the sample contains only zinc and its oxide, filter on a tared, sintered glass crucible, wash with cold water and finally with ether, dry in a vacuum desiccator, and weigh the zinc residue [9].

Many variations have been published. In one, the sample is boiled under reduced pressure at room temperature with 6–12% ammonium acetate, and filtered in an atmosphere of nitrogen [10].

Another advocates a solvent of 57·5 g ammonium chloride and 570 ml of ammonium hydroxide per litre [11].

In one publication it is recommended to stir a 1-g sample for 20 minutes at room temperature in a solution containing 2 g of ammonium chloride and 4 ml of ammonium hydroxide [12]. These reagent quantities for a 1-g sample are, in the author's opinion, a little too low. For most zinc powders or dusts, a 1-g sample should be extracted in not less than 50 ml of a solution containing 5–6 g of ammonium chloride and 5–10 ml of ammonium hydroxide.

One study has indicated that the optimum conditions for the determination of zinc oxide in the presence of metallic zinc demand, for 0·3–0·4 g of a sample, 94 ml of 5% ammonium acetate, 6 ml of 7M sodium hydroxide, and 1 g of calcium chloride; the shaking period was 10 minutes [13].

Other workers have recommended a simple 10-minute leach at room temperature with 100 ml of 20% ammonium chloride [14].

Some investigators have used quite different methods from the usual solvent action of ammonium acetate or ammoniacal ammonium chloride on zinc oxide. One suggests a solution of hot methanol and ricinoleic acid [15].

In another study, a solution of Trilon was used to extract oxides and basic oxides of zinc; sulphides are not affected [16].

A publication describes a method for dissolving the sample in an anhydrous solution of hydrochloric acid in acetic acid, and determining the water produced by titrating with a solution of iodine and sulphur dioxide in pyridine and methanol [17].

In another procedure, the sample is placed in a copper boat, wrapped in copper foil, and heated in a vacuum for 30 minutes at 450°C. This volatilizes zinc, leaving zinc oxide which can be brushed out of the boat and weighed [18].

One publication suggests the following technique. Pass hydrogen gas for 3 minutes into an extraction solution of 5 ml mercury plus 50 ml of 1N ammonium hydroxide containing 53 g ammonium chloride and 0·1 g of gelatin per litre, in a rubber-stoppered Erlenmeyer flask to expel air. Add quickly 1 g of the sample of zinc powder, pass in hydrogen for 1–2 minutes, and shake the flask to dissolve zinc oxide. Zinc amalgam in mercury is not dissolved in the extracting solution [19].

A determination of surface zinc oxide on zinc sulphide phosphors has been outlined. A weighed sample is slurried with water, and titrated to the methyl orange endpoint with 0·1N hydrochloric acid [20].

2. Metallic Zinc in Zinc Dust, Zinc Oxide, etc.

A. Ferric Sulphate Method

Metallic zinc reduces ferric iron to the ferrous state, whereas zinc oxide does not. The ferrous solution can be titrated with a standard solution of potassium permanganate or dichromate to give a measure of the metallic zinc present. This method has been in use for over seventy years; typical details are outlined below.

To a dry 500-ml graduated flask add 7 g of pure powdered ferric sulphate and 0·5 g of zinc dust; agitate the flask until they are thoroughly mixed. Add 25 ml water, stopper the flask, and shake for about 15 minutes, by which time the zinc dust should be dissolved. Add 300 ml of 1:1 sulphuric acid, and make the solution up to the mark with water. Withdraw 100 ml of the solution and titrate the ferrous sulphate, which has been formed, with 0·1N potassium permanganate. One ml of 0·1N $KMnO_4$ = 0·003269 g zinc. If a significant quantity of iron is present in the sample, dissolve 1 g in sulphuric acid, determine the number of ml of 0·1N potassium permanganate required to oxidize it, and deduct a proportionate amount from the previous titration. Potassium dichromate, with sodium diphenylamine sulphonate indicator, may be substituted for potassium permanganate [21].

Several variations of this procedure, and comments on its validity, have been published. The influence of metallic lead on the titration of ferrous iron has been found negligible [22].

If zinc sulphide is present, it will reduce ferric iron. The effect of natural zinc sulphide and of zinc sulphide separated from incrustations on molten metal, as well as that of lead sulphide, is considerably less than that of zinc sulphide precipitated from aqueous solution. In fact, the reduction of iron is only about 6–8% calculated as metallic zinc [23].

The use of ferric ammonium sulphate and an acetate buffer has been proposed [24].

In another study, the optimum concentrations of ferric sulphate or ferric ammonium sulphate, $NH_4Fe(SO_4)_2.12H_2O$, were stated to be 250 and 600 g/litre, respectively [25].

Traces of zinc in zinc oxide have been measured colorimetrically after the reduction of ferric to ferrous iron in 10% sulphuric acid containing 1,10-phenanthroline; the sample is kept cold for 4 hours with the aid of dry ice. Addition of ammonium bifluoride removed the interference of the remaining excess ferric iron, and the absorbance of the complex with ferrous iron was measured at 510 nm [26].

B. Hydrogen Evolution

Metallic zinc in zinc dust may be determined by the volume of hydrogen evolved when the sample is dissolved in 1:1 sulphuric or hydrochloric acid. This method has been used successfully for over fifty years; a representative outline is given below.

Into a small Erlenmeyer flask is placed 1 g of zinc dust, together with a small piece of sheet platinum and about 5 g of ferrous sulphate. The flask is nearly filled with distilled water saturated at room temperature with hydrogen gas. The platinum and ferrous sulphate exert a catalytic effect on hydrogen evolution; the ferrous sulphate also coagulates the zinc dust when it becomes wetted and tends to minimize the floating of these particles.

A rubber stopper inserted into the neck of the Erlenmeyer flask contains a small separatory funnel and a connecting tube provided with a three-way stopcock. The connecting tube is joined to a gas measuring tube, which in turn is connected with a levelling bottle containing 10% sulphuric acid saturated with hydrogen at room temperature. Water is now run in from the separatory funnel to displace all the air in the flask and the connecting tube. The stopcock in the connecting tube is turned so that the downward inlet is in connexion with the measuring tube. By raising the levelling bottle all the gas in the measuring tube is displaced. The stopcock is now turned through 90 degrees to connect the decomposing flask with the measuring tube.

About 30 ml of 1:1 sulphuric acid are poured into the separatory funnel. A small portion of this acid is allowed to run into the decomposing flask until a steady evolution of hydrogen occurs. The gas evolved, together with some solution and a very small amount of zinc dust passes over into the measuring tube, displacing the acid there. During this time the acid in the measuring tube and flask is shaken to wash down the particles of zinc dust from the upper parts of the tube and flask now filled with gas. The particles in the measuring tube, on coming in contact with the 10% sulphuric acid, are readily dissolved and generate their portion of hydrogen.

When all the zinc dust has been dissolved, water is run in from the separatory funnel to force the hydrogen over into the measuring tube, and to fill the flask and connecting tube with water through the stopcock, which is then closed. After levelling with the levelling bottle, the volume of hydrogen generated from a 1-g sample at the prevailing atmospheric conditions is read from the measuring tube. The percentage of metallic zinc is then calculated from the following expression:

$$\%\ \text{metallic zinc} = \frac{V \times (P - p - a) \times 0{\cdot}29183}{(1 + 0{\cdot}00367t)\,760}$$

in which V = volume of gas in the measuring tube at atmospheric conditions, P = barometric pressure, p = vapour pressure of water above 10% sulphuric acid at room temperature, a = temperature correction for expansion of mercury in the barometer, which is approximately 0·13t, and t = room temperature in °C.

It must be remembered, of course, that any other metals present which evolve hydrogen with sulphuric acid will interfere. This gasometric method, which requires about 1·5 hours, is capable of giving very satisfactory results [27, 28].

Slight variations of this basic procedure have been mentioned in other publications [29, 30]. The determination has been made in the presence of zinc sulphide, in a vacuum apparatus [31].

C. Miscellaneous Methods

A number of other methods for zinc in the presence of zinc oxide have been published.

When zinc dust is treated with a solution containing 18·5 g/litre of mercuric chloride, which has been adjusted to *p*H 6 with a solution of 80 g/litre of potassium cyanide, metallic zinc is dissolved and zinc oxide remains as a residue [32].

Metallic zinc can be separated from traces of oxide by dissolution in mercury, which does not react with zinc oxide. The low content of the latter can be determined by a coulometric method, and metallic zinc by difference [33].

Zinc metal has been determined by acidifying with hydrochloric acid a 0·5 g sample containing 2–5 ml of 0·4–0·45N potassium dichromate and 2 g potassium iodide, and titrating the excess of the oxidant [34].

Another method is based on the concentration of reducing agents present before and after the removal of metallic zinc as zinc chloride. To a 10-g sample add 25 ml of 0·1N iodine, mix, and allow to stand for 1 hour in the dark. Add 100 ml of 10% hydrochloric acid, dissolve, and titrate with standard sodium thiosulphate solution. Dissolve a second 10-g sample in 100 ml of 10% hydrochloric acid, allow to stand for 1 hour, add 25 ml of 0·1N iodine solution, mix, allow to stand in the dark 30 minutes, and titrate with standard sodium thiosulphate [35].

An iodimetric determination of excess zinc in zinc oxide has been published. Treat with hot concentrated sulphuric acid an excess of iodine, cool, decant, and dilute to obtain 0·5N sulphuric acid. Take 5 ml of this solution, and titrate with 0·005N sodium thiosulphate in an inert atmosphere, adding near the endpoint potassium iodide and starch. Dissolve a 0·2-g sample in 5 ml of the solution as above and titrate. The difference in sodium thiosulphate used is a measure of the amount of iodine which reacted with the excess of zinc present [36].

In another proposal, the sample was stirred for 1 minute with 50 ml of 1·5% copper sulphate solution, and 10 ml of M ferric sulphate solution added. The mixture was acidified with phosphoric acid, and titrated with 0·1N potassium permanganate [37].

A rapid colorimetric method has been described. The sample is mixed with 0·001N potassium permanganate solution which has been mixed with chloride-free sulphuric acid in the ratio of 10:1. After dissolution of the sample, the permanganate content is determined in a colorimeter [38].

Another colorimetric procedure has been published. Prepare a reference solution from 20 ml of an acid solution containing 65 ml sulphuric acid and 65 ml phosphoric acid diluted to 1 litre, 25 ml of water, 5 g of zinc oxide, 25 ml of 0·0001 N potassium dichromate, 90 ml of water, and 5 ml of 25% solution of diphenylcarbazide in acetone, in that order. Prepare the sample solution from the same reagents, but add potassium dichromate first. Prepare a blank with all reagents except potassium dichromate. Measure the absorbance at 520 nm [39, 40].

Compounds of zinc have been subjected to the following phase analyses:

(a) Digestion with ammoniacal Trilon solution containing sodium dibutyl naphthalene sulphonate as an inhibitor, to dissolve oxidized zinc compounds.
(b) Treatment with copper nitrate solution, and determination of metallic zinc in the filtrate.
(c) Leaching with Trilon and hydrogen peroxide to dissolve zinc sulphide.
(d) Treatment of the final residue with hydrochloric acid to dissolve zinc stannate [41, 42].

3. Zinc Minerals

Several zinc minerals have been differentiated by chemical phase analyses. In the discussion of zinc oxide earlier in this chapter, the differentiation of the oxidized zinc minerals willemite, hemimorphite, hopeite, parahopeite, tarbuttite, zincite, descloizite, and smithsonite from the sulphide sphalerite was described [8].

Zinc silicate can be selectively dissolved in 0·1N sulphuric acid at room temperature for 2 hours, and zinc ferrite in the precipitate can be dissolved in dilute hydrochloric acid, leaving zinc aluminate in a final residue [15].

Oxidized zinc minerals, zinc sulphide, and zinc stannate may be separated by the treatment scheme outlined earlier at the end of the preceding section—**2**.C. Miscellaneous Methods [41, 42].

A phase analysis for calamine, sphalerite, adamite, descloizite, and smithsonite has been published [43]. A differentiation of calamine and sphalerite in the presence of adamite, descloizite, and smithsonite is difficult, because the selective tartaric acid solvent for calamine and the acidified ferric chloride solvent for sphalerite dissolve the zinc of adamite and some of the zinc of descloizite. In the absence of smithsonite, 5% copper sulphate solution is a selective solvent for calamine. Since descloizite dissolves in dilute hydrochloric acid in which sphalerite is practically completely insoluble, samples containing descloizite are heated with 1% hydrochloric acid before treatment with ferric chloride solution. By the use of a 5% copper sulphate solution, tartaric acid and an ammonium chloride/ammonium acetate solution, it is possible to determine the zinc of calamine and of adamite separately.

It has been stated by the same worker that the best solvent for descloizite is 1% hydrochloric acid. For a –200 mesh sample, 1-hour contact is sufficient [44].

It has been recommended to separate smithsonite from sphalerite by dissolving the former in a solution of 25% ammonium hydroxide and 25% ammonium chloride [45].

Another study indicated that zinc silicate dissolves in 2% tartaric acid, smithsonite in ammoniacal ammonium chloride solution, and sphalerite in acidified ferric chloride solution [46].

One worker has reported that treatment of a mixed willemite–sphalerite 0·3-g sample for 1 hour with 50 ml of 5% acetic acid dissolved the zinc silicate mineral but had only slight effect on sphalerite [47].

REFERENCES

1. Young, R. S., *Chemical Analysis in Extractive Metallurgy*, London, Charles Griffin, 1971.
2. Abdeev, M. A., and Trofimova, S. G., *Trudy Altaisk. Gorno-Met. Nauch.-Issledovatel. Inst., Akad. Nauk Kazakh. S.S.R.* **1**, 67–78 (1954). *C.A.* **51**, 14481 (1957).
3. Kitagawa, H., and Shibata, N., *Nippon Kinzoku Gakkaishi* **26**, 186–90 (1962). *C.A.* **61**, 3695 (1964).
4. Mayants, A. D., *Zavod. Lab.* **13**, 920–3 (1947). *C.A.* **44**, 3406 (1950).
5. Pellowe, E. F., and Hardy, F. R. F., *Analyst* **77**, 208–10 (1952).
6. Sudilovskaya, E. M., and Filippova, N. A., *Analiz Rud Tsvetnykh Metal. i Produktov ikh Pererabotki, Sbornik Nauch. Trudov* No. **14**, 138–42 (1958). *C.A.* **53**, 13887 (1959).
7. Trofimova, S. G., and Abdeev, M. A., *Zavod. Lab.* **25**, 1443–5 (1959). *C.A.* **54**, 8439 (1960).
8. Barker, C. W., and Young, R. S., *J. Soc. Chem. Ind.* **67**, 61 (1948).
9. Osborn, G. H., *Analyst* **76**, 114–15 (1951).
10. Taylor, C. G., *Analyst* **83**, 425–9 (1958).
11. Zinov'eva, L. D., and Bitarova, V. A., *Metallurg. i Khim. Prom. Kazakhstana, Nauchno-Tekhn. Sb.* **1962**, (6), 71–3. *C.A.* **61**, 1274 (1964).
12. Imai, H., *J. Chem. Soc. Japan, Pure Chem. Sect.* **74**, 834–7 (1953). *C.A.* **48**, 4366 (1954).
13. Ospanov, K. K., and Alimpeva, S. D., *Zavod. Lab.* **37**, 1051–2 (1971). *C.A.* **76**, 30382 (1972).

14. KHRISTOFOROV, B. S., and FOMINYKH, V. S., *Fazovyi Khim. Analiz. Rud i Mineralov. Lening. Gos. Univ.* **1962**, 95–9. *C.A.* **58**, 11955 (1963).
15. FRENAY, E., COLLEÉ, R., and GRODENT, H., *Anal. Chim. Acta* **6**, 31–41 (1952). *C.A.* **46**, 7469 (1952).
16. FILIPPOVA, N. A., and KOROSTELEVA, V. A., *Zavod. Lab.* **25**, 535–9 (1959). *C.A.* **53**, 16805 (1959).
17. EBERIUS, E., and KOWALSKI, W., *Z. Erzbergbau u Metallhuttenw.* **7**, 229–34 (1954). *C.A.* **51**, 13650 (1957).
18. BALIS, E. W., BRONK, L. B., and LIEBHAFSKY, H. A., *Anal. Chem.* **21**, 1373–4 (1949).
19. YOSHIDA, M., and ARATA, M., *Bunseki Kagaku* **15**, 441–5 (1966). *C.A.* **65**, 17688 (1966).
20. LARACH, S., and THOMSEN, S. M., *Anal. Chem.* **26**, 1600–1 (1954).
21. DILLON, V. S., *Assay Practice on the Witwatersrand, Johannesburg, Transvaal and Orange Free State Chamber of Mines*, 1955.
22. KHRISTOFOROV, B. S., and ARTEMENKO, A. R., *Sbornik Nauch. Trudov Vsesoyuz. Nauch.-Issledovatel. Gorno-Met. Inst. Tsvetnoi Met.* **1958**, No. 3, 282–4. *C.A.* **53**, 21412 (1959).
23. KHRISTOFOROV, B. S., ARTEMENKO, A. R., and BITAROVA, V. A. *Sbornik Trudov Vsesoyuz. Nauch.-Issledovatel, Gorno-Met. Inst Tsvetnoi Met.* **1959**, No. 5, 151–5. *C.A.* **55**, 26838 (1961).
24. EBERIUS, E., *Z. Erzbergbau Metallhuttenw.* **17**, 181–91 (1964). *C.A.* **61**, 3690 (1964).
25. RACZKA, E., *Rudy Metale Niezelaz.* **13**, 526–8 (1968). *C.A.* **70**, 64016 (1969).
26. KRUSE, J. M., *Anal. Chem.* **43**, 771–3 (1971).
27. COPEMAN, D. A., *J. South African Chem. Institute* **XXV**, No. 2, 62–6 (1942).
28. FURMAN, N. H., ed., *Scott's Standard Methods of Chemical Analysis*, 6th ed., Vol. 1, Princeton, N.J., Van Nostrand, 1962.
29. ALLSOPP, H. J., *Analyst* **82**, 474–83 (1957).
30. LINDLEY, G., *Lab. Practice* **10**, 628–9 (1961).
31. GROMOV, L. A., and OSIPOV, V. A., *Zh. Analit. Khim.* **19**, 189–94 (1964). *C.A.* **60**, 15138 (1964).
32. EK, C., *Rev. universelle Mines* **13**, 249–53 (1957). *C.A.* **54**, 7434 (1960).
33. SALATA, A., and ZANAROLI, L., *Met. Ital.* **60**, 486–90 (1968). *C.A.* **69**, 92725 (1968).
34. IMAI, H., and SUEHIRO, S., *J. Chem. Soc. Japan, Pure Chem. Sect.* **74**, 694–7 (1953). *C.A.* **48**, 3195 (1954).
35. GORELIK, D. S., and KAZAKOVA, O. A., *Lakokrasochnye Materialy i ikh Primenenie* **1963**, No. 5, 50–1. *C.A.* **60**, 2325 (1964).
36. DEREN, J., and FRYT, E., *Chem. Anal.* **8**, 365–7 (1963). *C.A.* **60**, 2325 (1964).
37. HAHN, F. L., *Z. Anal. Chem.* **174**, 261–2 (1960). *C.A.* **55**, 2352 (1961).
38. DEREN, J., and KOWALSKA, A., *Chem. Anal.* **7**, 563–6 (1962). *C.A.* **57**. 15802 (1962).
39. NORMAN, V. J., *Analyst* **89**, 261–5 (1964).
40. NORMAN, V. J., *Analyst* **97**, 156–7 (1972).

41. Filippova, N. A., and Korosteleva, V. A., *Zavod. Lab.* **25**, 1053–9 (1959). *C.A.* **54**, 8430 (1960).
42. Filippova, N. A., Martynova, L. A., Selezneva, M. N., and Stepareva, V. N., *Sb. Nauch. Tr., Nauch.-Issled. Inst. Tsvet. Metal.* **34**, 179–82 (1971). *C.A.* **77**, 172271 (1972).
43. Leont'eva, K. D., *Analiz Rud Tsvetnykh Metal. i Produktov ikh Pererabotki, Sbornik Nauch. Trudov* **1958**, No. 14, 93–102. *C.A.* **53**, 13871 (1959).
44. Leont'eva, K. D., *Sbornik Nauch. Trudov Gosudarst. Nauch.-Issledovatel. Inst. Tsvetnoi Met.* **1958**, No. 14, 93–102. *C.A.* **54**, 10652 (1960).
45. Ionescu, M., and Pavel, R., *Rev. minelor* **9**, 39–44 (1958). *C.A.* **53**, 12092 (1959).
46. Bayula, A. G., and Alekhina, K. N., *Soobshchen. Dal'nevostoch. Filiala Sibir. Otdel. Akad. Nauk S.S.S.R.* **1959**, 265–8. *C.A.* **55**, 12147 (1961).
47. Babcan, J., *Sb. Geol. Ved, Technol. Geochemie* No. **3**, 27–32 (1964). *C.A.* **61**, 1273 (1964).

Zirconium

A few studies have been made on the phase analysis of zirconium compounds.

Metallic zirconium has been separated from its oxide by the solubility of the former in cold, dilute hydrofluoric acid [1].

In another method, metallic zirconium is determined from the volume of hydrogen liberated when the sample is dissolved in hydrofluoric acid [2].

A differentiation of zirconium metal, carbide and oxide has been reported. Digestion with 1:20 hydrofluoric acid, and filtration, will put metallic zirconium into the filtrate. Digestion of the residue with 1:2 sulphuric acid, and filtration, will give zirconium carbide in the filtrate, leaving the oxide in the final residue [3].

A phase analysis of zirconium diboride has been published [4].

REFERENCES

1. Wada, I., and Ishii, R., *Bull. Inst. Phys. Chem. Research* **21**, 877–83 (1942). *C.A.* **42**, 8705 (1948).
2. Straumanis, M. E., and Ejima, T., *Z. Anal. Chem.* **177**, 241–4 (1960).
3. Kozyreva, L. S., Kuteinikov, A. F., and Zharova, N. P., *Zavod. Lab.* **30**, 1328–9 (1964). *C.A.* **62**, 3400 (1965).
4. Bogova, L. V., *Tr. Vses. Inst. Nauch.-Issled. Proekt. Rab. Ogneupor. Prom.* No. **37**, 164–78 (1965). *C.A.* **66**, 121739 (1967).

Author Index

Subject Index